T0214245

Communications in Computer and Information Science 1196

More information about this series at http://www.springer.com/series/7899

Bart Bogaerts · Gianluca Bontempi ·
Pierre Geurts · Nick Harley ·
Bertrand Lebichot · Tom Lenaerts ·
Gilles Louppe (Eds.)

Artificial Intelligence and Machine Learning

31st Benelux AI Conference, BNAIC 2019,
and 28th Belgian-Dutch Machine Learning Conference, BENELEARN 2019
Brussels, Belgium, November 6–8, 2019
Revised Selected Papers

 Springer

Editors
Bart Bogaerts 🆔
Vrije Universiteit Brussel
Brussels, Belgium

Gianluca Bontempi 🆔
Université Libre de Bruxelles
Brussels, Belgium

Pierre Geurts 🆔
Université de Liège
Liège, Belgium

Nick Harley
Vrije Universiteit Brussel
Brussels, Belgium

Bertrand Lebichot 🆔
Université Libre de Bruxelles
Brussels, Belgium

Tom Lenaerts 🆔
Université Libre de Bruxelles
Brussels, Belgium

Gilles Louppe 🆔
Université de Liège
Liège, Belgium

ISSN 1865-0929   ISSN 1865-0937 (electronic)
Communications in Computer and Information Science
ISBN 978-3-030-65153-4   ISBN 978-3-030-65154-1 (eBook)
https://doi.org/10.1007/978-3-030-65154-1

This Springer imprint is published by the registered company Springer Nature Switzerland AG
The registered company address is: Gewerbestrasse 11, 6330 Cham, Switzerland

Preface

The 31st edition of the annual Benelux Conference on Artificial Intelligence (BNAIC 2019) and the 28th Belgian Dutch Conference on Machine Learning (Benelearn 2019) were jointly organized by the Vrije Universiteit Brussel, the Université Libre de Bruxelles, and the Université de Liège, under the auspices of the Benelux Association for Artificial Intelligence (BNVKI) and the Dutch Research School for Information and Knowledge Systems (SIKS), in collaboration with Brewery of Ideas under the name AI Synergies.

Held yearly, the objective of these conferences is to promote and disseminate recent research developments in Artificial Intelligence (AI) and Machine Learning (ML) in the Benelux. For the first time, we have combined these two conferences with a whole-day business track (on November 6, 2019) and an 'Art-AI' evening (also November 6, 2019) into an overarching event called AI Synergies, with the goal of further expanding the reach and impact of the two conferences and to allow people from industry to get in touch with the academics and learn about their work. This part of the conference was co-organized by the company Brewery of Ideas.

For the scientific part, we welcomed four types of contributions, namely A) regular papers, B) compressed contributions, C) demonstration abstracts, and D) bachelor/master thesis abstracts. We received 127 submissions, 66 for the AI and 61 for the ML tracks. In terms of the four types, we received 10 demonstration abstracts, 24 bachelor/master thesis's, 50 regular paper submissions, and 43 compressed contributions. Submissions were evaluated by three reviewers. From all these submissions, the International Program Committee (see at end) selected 115 for the conference, to be presented either as talk or research pitches in sessions organized around the most important AI and ML topics.

Finally, the BNAIC/Benelearn Program Committee selected the most insightful and relevant papers from both the ML and AI tracks to be included in these post-proceedings. An additional peer-review process was single blind with three reviewers per paper. In total 11 papers were withheld, resulting in the post-proceedings you are currently reading.

For the AI part, the selected papers reflect the diversity of submissions that were also presented at the conference. The post-proceedings include two contributions on the topic of modular design of robot swarms and one contribution that aims to help in the fight against HIV. Additionally, the AI part has a contribution on computational models that can help understand how humans perceive language, as well as a novel method for exploring latent spaces of datasets.

For the ML part, the ratio between fundamental and applied papers is quite balanced: three for each. The applied part of the ML post-proceedings includes papers on various fields: churn prediction in telecommunications, a recommendation system, and a data visualization tool. On the fundamental side, the papers cover defense against adversarial attacks, ordinal classification problems, and detrimental point processes.

With the booming interest in AI and ML in both academia and companies, this edition of the BNAIC/Benelearn conferences was an outstanding success. It revealed that both communities are active and strong in the Benelux, and we hope that future organizations of this event will surpass the one organized in 2019 in Brussels.

July 2020

Bart Bogaerts
Gianluca Bontempi
Pierre Geurts
Nick Harley
Bertrand Lebichot
Tom Lenaerts
Gilles Louppe

Organization

General Chairs

Tom Lenaerts	Université Libre de Bruxelles, Belgium
Johan Loeckx	Vrije Universiteit Brussel, Belgium
Claudio Truzzi	iCity and Université Libre de Bruxelles, Belgium

Program Committee Chairs

Bart Bogaerts	Vrije Universiteit Brussel, Belgium
Gianluca Bontempi	Université Libre de Bruxelles, Belgium
Pierre Geurts	Université de Liège, Belgium
Nick Harley	Vrije Universiteit Brussel, Belgium
Bertrand Lebichot	Université Libre de Bruxelles, Belgium
Gilles Louppe	Université de Liège, Belgium

Industry Chairs

Peter Buelens	Brewery of Ideas, Belgium
Jonathan Duplicy	Innoviris, Belgium
Annerieke Heuvelink	Philips, The Netherlands
Johan Loeckx	Vrije Universiteit Brussel, Belgium
Leander Schietgat	Vrije Universiteit Brussel, Belgium
Claudio Truzzi	iCity and Université Libre de Bruxelles, Belgium

Demo Chairs

Katrien Beuls	Vrije Universiteit Brussel, Belgium
Paul Van Eecke	Vrije Universiteit Brussel, Belgium

Arts Chairs

Christophe De Jaeger	Gluon, Belgium
Luc Steels	Vrije Universiteit Brussel, Belgium
Geraint Wiggins	Vrije Universiteit Brussel, Belgium

Program Committee

Martin Atzmueller	Tilburg University, The Netherlands
Reyhan Aydogan	Technische Universiteit Delft, The Netherlands
Souhaib Ben Taieb	Université de Mons, Belgium
Floris Bex	Universiteit Utrecht, The Netherlands

Evgueni Smirnov	Universiteit Maastricht, The Netherlands
Felix Sommer	Université catholique de Louvain, Belgium
Gerasimos Spanakis	Universiteit Maastricht, The Netherlands
Jennifer Spenader	University of Groningen, The Netherlands
Miguel Suau	Technische Universiteit Delft, The Netherlands
Johan Suykens	Katholieke Universiteit Leuven, Belgium
Dirk Thierens	Universiteit Utrecht, The Netherlands
Frank Van Harmelen	Vrije Universiteit Amsterdam, The Netherlands
Remco Veltkamp	Universiteit Utrecht, The Netherlands
Joost Vennekens	Katholieke Universiteit Leuven, Belgium
Arnoud Visser	Universiteit van Amsterdam, The Netherlands
Willem Waegeman	University of Groningen, Belgium
Louis Wehenkel	Université de Liège, Belgium
Gerhard Weiss	Universiteit Maastricht, The Netherlands
Wolfgang Wiedermann	University of Missouri, USA
Marco Wiering	Rijksuniversiteit Groningen, The Netherlands
Jef Wijsen	Université de Mons, Belgium
Mark H. M. Winands	Universiteit Maastricht, The Netherlands
Marcel Worring	Universiteit van Amsterdam, The Netherlands
Duo Xu	Georgia Institute of Technology, USA
Yingqian Zhang	Technische Universiteit Eindhoven, The Netherlands
Tim van Erven	Universiteit Leiden, The Netherlands
Nanne van Noord	Universiteit van Amsterdam, The Netherlands
Marieke van Vugt	Rijksuniversiteit Groningen, The Netherlands
Egon L. van den Broek	Universiteit Utrecht, The Netherlands
Peter van der Putten	Universiteit Leiden, The Netherlands
Leon van der Torre	University of Luxembourg, Luxembourg

Additional Reviewers

Wissam Siblini
Menno van Zaanen

Contents

Artificial Intelligence Part: BNAIC

AutoMoDe-IcePop: Automatic Modular Design of Control Software
for Robot Swarms Using Simulated Annealing   3
  Jonas Kuckling, Keneth Ubeda Arriaza, and Mauro Birattari

Evaluation of Alternative Exploration Schemes in the Automatic Modular
Design of Robot Swarms   18
  *Gaëtan Spaey, Miquel Kegeleirs, David Garzón Ramos,
  and Mauro Birattari*

Towards a Phylogenetic Measure to Quantify HIV Incidence   34
  *Pieter Libin, Nassim Versbraegen, Ana B. Abecasis, Perpetua Gomes,
  Tom Lenaerts, and Ann Nowé*

Cognitively Plausible Computational Models of Lexical Processing
Can Explain Variance in Human Word Predictions and Reading Times   51
  Wietse de Vries

Latent Space Exploration Using Generative Kernel PCA   70
  David Winant, Joachim Schreurs, and Johan A. K. Suykens

Machine Learning Part: Benelearn

Calibrated Multi-probabilistic Prediction as a Defense Against
Adversarial Attacks   85
  Jonathan Peck, Bart Goossens, and Yvan Saeys

Machine Learning Methods for Ordinal Classification with Additional
Relative Information 126
  Mengzi Tang, Raúl Pérez-Fernández, and Bernard De Baets

Towards Deterministic Diverse Subset Sampling 137
  J. Schreurs, M. Fanuel, and J. A. K. Suykens

Extended Bayesian Personalized Ranking Based
on Consumption Behavior 152
  Alireza Gharahighehi and Celine Vens

SubSect—An Interactive Itemset Visualization 165
  Joey De Pauw, Sandy Moens, and Bart Goethals

Understanding Telecom Customer Churn with Machine Learning:
From Prediction to Causal Inference 182
    Théo Verhelst, Olivier Caelen, Jean-Christophe Dewitte,
    Bertrand Lebichot, and Gianluca Bontempi

Author Index 201

Artificial Intelligence Part: BNAIC

AutoMoDe-IcePop: Automatic Modular Design of Control Software for Robot Swarms Using Simulated Annealing

Jonas Kuckling[(✉)], Keneth Ubeda Arriaza, and Mauro Birattari

IRIDIA, Université Libre de Bruxelles, Brussels, Belgium
{jonas.kuckling,mbiro}@ulb.ac.be

Abstract. Prior research has shown that the optimization algorithm is an integral part of the automatic modular off-line design of control software for robot swarms and can have great influence on the quality of the control software produced. In this paper we investigate, whether a stochastic local search metaheuristic—simulated annealing—can be used as the optimization algorithm in the automatic modular design of robot swarms. The results indicate that simulated annealing is indeed a viable candidate. Additionally, we investigate the influence of some obvious variations of simulated annealing on the performance of the automatic modular design.

Keywords: Swarm robotics · Automatic design · Simulated annealing

1 Introduction

Designing control software for a robot swarm is a challenging task, as the global desired behavior usually emerges from the interactions of the robots between each other and the environment [10,37]. Manual software design therefore often relies on trial-and-error [4] and a general methodology for designing control software for robot swarms is still missing [12].

Automatic design offers a promising alternative, by transforming the design problem into an optimization problem. Instead of writing control software that performs a specific mission, a target architecture is optimized with regard to a mission-dependent objective function. A popular automatic design approach is neuro-evolutionary swarm robotics which uses evolutionary algorithms to design artificial neural networks. While this approach has successfully been applied to many missions [8,11,21,33,35,36], multiple challenges remain to be solved [5,31,34]. The most important is the weak transferability of the generated control software, resulting in performance drops when deployed in reality. This drop

---

JK and KUA contributed equally to this work and should be considered as co-first authors. The experiments were designed by JK and performed by KUA. The paper was drafted by JK and edited by MB; all authors read and commented the final version. The research was directed by MB.

© Springer Nature Switzerland AG 2020
B. Bogaerts et al. (Eds.): BNAIC 2019/BENELEARN 2019, CCIS 1196, pp. 3–17, 2020.
https://doi.org/10.1007/978-3-030-65154-1_1

in performance is often associated with the reality gap—inherent differences between the design context of the simulation and the real world.

Francesca et al. [14] see in this phenomenon a resemblance to the problem of over-fitting in machine learning. Analogous to the bias-variance trade-off [9,17], they propose to introduce a bias to the automatic design process. Their proposed bias is a restriction of possible control software, by defining a control architecture which can be composed through the combination of predefined modules. As a proof of concept, Francesca et al. implemented AutoMoDe-Vanilla, an automatic modular design approach that generates finite-state machines with up to four states. Such generated finite-state machines are composed of states, which will execute an associated behavior as long as they are active, and transitions, that have an associated probabilistic condition which can trigger the transition from one state to another. Vanilla uses F-race [2] to combine the finite-state machines out of a set of predefined modules (behaviors and conditions) and to fine tune their parameters.

With AutoMoDe-Chocolate [13], Francesca et al. implemented a variant of Vanilla that differs only in the optimization algorithm employed. Chocolate uses Iterated F-race [3], instead of F-race. The results of their experiments show that Chocolate performs significantly better than Vanilla on many missions. Given that the only difference between the two methods is the optimization algorithm it seems apparent that the optimization algorithm is an important part of the automatic modular design approach and can have a great influence on the performance of generated control software. Following up on this observation, we create IcePop, another instance of AutoMoDe. It is functionally similar to Chocolate and Vanilla but it uses simulated annealing as an optimization algorithm. We choose simulated annealing because it is a well-studied algorithm [6,19,26,29,32] that has found many applications (for surveys see for example [1] and [32]).

Simulated annealing is a metaheuristic inspired by the thermodynamical process of annealing [23]. At higher temperatures the particles in a crystal are more excited and can move more freely than at lower temperatures. Similarly, the simulated annealing algorithm has a "temperature" parameter. When it is high, the algorithm has a chance to accept worsening solutions, mimicking the free movement of the particles. At lower temperatures, the algorithm will select worsening solutions less likely, thus constraining the movement of the solution candidate. Simulated annealing has shown properties that are desirable for the automatic design of control software. It has been shown to effectively traverse the search space and to converge quickly towards promising solutions [22]. This allows an efficient use of the allocated budget. Furthermore, simulated annealing contains mechanisms to escape local optima—e.g., by accepting worsening moves at higher temperatures. Without any a priori knowledge of the shape of the search space, this is an important property as it reduces the risk of premature convergence to suboptimal solutions.

The rest of this paper is structured as follows: In Sect. 2 we present the experimental setup that we used—the robotic platform, the design methods and the experimental protocol. In Sect. 3 we present four experiments and their results. In Sect. 4 we summarize our findings and give an outlook to future work.

2   Experimental Setup

In this section we describe the experimental setup and protocol that was used to obtain the results described in Sect. 3.

2.1   Robotic Platform

`IcePop` designs control software for a swarm of modified e-puck robots [16,30]. The e-puck robots are equipped with two wheels, whose velocity can be adjusted independently, three ground sensors that can perceive the greyscale color value of the floor, and eight IR transceivers that are spaced equally around the robot, that can perceive proximity and light values. The robot is also equipped with a range-and-bearing board [18] that comprises twelve IR emitters and twelve receivers equally distributed along the perimeter of the board and pointed radially and outwards, on the horizontal plane. The range-and-bearing board allows the e-puck to send and receive messages within a range of 0.7m. In order to abstract the actual sensor configuration, we use a reference model [20]. Specifically, we use RM1.1 (see Table 1), the reference model that was used to define the modules of `Chocolate`.

In this reference model, each robot has eight light and proximity sensors returning floating point values between 0 and 1. $prox_i$ and $light_i$ define the proximity and light reading for the ith sensor respectively. Three ground sensors ($ground_i$) return one of three values, indicating whether the ground underneath them is black, gray or white. The reference model uses the range-and-bearing board to count the number of neighbors in communication range (n) and calculates an attraction vector (V_d) towards the center of mass of all perceived robots. Additionally the robot has two wheels, whose velocity can be adjusted independently (v_l and v_r for the velocity of the left wheel and the right wheel respectively).

2.2   Automatic Design Methods

We compare two automatic modular design methods: `Chocolate` and `IcePop`. `Chocolate` [13] generates probabilistic finite-state machines with up to four states. For that it uses a set of six behaviors and six conditions that are defined on top of RM1.1 [20]. The six behaviors are: exploration, stop, phototaxis, anti-phototaxis, attraction and repulsion. The six conditions are: black-floor, gray-floor, white-floor, neighbor-count, inverted-neighbor-count and fixed-probability. For a detailed description of the modules, we refer the reader to their original definition [14]. The optimization algorithm used by `Chocolate` is Iterated F-race [27].

Table 1. Reference model RM1.1 [20]. Sensors and actuators of the e-puck robot. The period of the control cycle is 100 ms.

Sensor/Actuator	Parameters	Values
proximity	$prox_i$, with $i \in \{0, \ldots, 7\}$	$[0, 1]$
light	$light_i$, with $i \in \{0, \ldots, 7\}$	$[0, 1]$
ground	$ground_i$, with $i \in \{0, \ldots, 2\}$	$\{black, gray, white\}$
range-and-bearing	n	$\{0, \ldots, 19\}$
	V_d	$([0, 0.7]m, [0, 2\pi]$ radian$)$
wheels	v_l, v_r	$[-0.12, 0.12]$ m/s

Algorithm 1. Component-based simulated annealing algorithm

---

best solution $s^* :=$ incumbent solution $\hat{s} := s_0$
$i := 0$
$T_0 :=$ initialize temperature according to *initial temperature*
while *stopping criterion* is not met **do**
    choose a solution s_{i+1} in the *neighborhood* of $\hat{s}$ according to *exploration criterion*
    if s_{i+1} meets acceptance criterion **then**
        $\hat{s} := s_{i+1}$
        if $\hat{s}$ improves over s^* **then**
            $s^* := \hat{s}$
        end if
    end if
    if temperature length steps since last temperature update **then**
        update temperature according to *cooling scheme*;
    end if
    reset temperature according to *temperature restart mechanism*;
    $i := i + 1$
end while
return s^*

In this paper, we propose IcePop. It is based on Chocolate, as it uses the same modules and target architecture. The difference between the two methods is that IcePop adopts the component-based simulated annealing algorithm (see Algorithm 1) as the optimization algorithm. Franzin and Stützle proposed this component-based algorithm in an effort to unify many variants of the simulated annealing algorithm [15]. We choose to adopt this algorithm because it provides the flexibility to easily change components should the need arise.

The component-based simulated annealing algorithm contains placeholders for commonly used components. In Table 2, we present our choices of components that we use in the implementation of the simulated annealing for IcePop. The initial solution supplied to the algorithm is a minimal valid instance of control software. In our case this is a finite-state machine with exactly one state executing the stop behavior. The neighborhood function is implicitly defined through the application of a random valid perturbation operator. In IcePop, we

Table 2. Configuration of the simulated annealing algorithm.

Component	Type	Parameter
Initial solution	Minimal controller	Stop behavior
Neighborhood	Defined through perturbation operators	
Initial temperature	Fixed value	$T_0 = 125.0$
Stopping criterion	Budget of simulations	50000 simulations
Exploration criterion	Random exploration	Valid perturbation operators
Acceptance criterion	Metropolis condition	Mean with 10 samples
Temperature length	Fixed value	$T_{length} = 1$
Cooling scheme	Geometric cooling	$\alpha = 0.9782$
Temperature restart	Fixed value	Every 5000 simulations

have defined eleven perturbation operators: adding a state, removing a state, adding a transition, removing a transition, changing the initial state, changing the starting point of a transition, changing the end point of a transition, changing the behavior associated with a state, changing the condition associated with a transition, changing the parameters of a behavior, and changing the parameters of a condition. The initial temperature is set to 125.0. The stopping criterion is defined as a maximum budget of simulation runs. That is, after the allocated budget of simulation runs is exhausted, the algorithm should return the final instance of control software. The exploration criterion selects a random valid perturbation operator and applies it on the incumbent solution. The acceptance criterion is the Metropolis condition [23, 28] that accepts or rejects new solutions based on their performance. The Metropolis condition will always accept an improving solution, and will accept a worsening solution with probability $\exp(-(e - e')/T)$ where T is the current temperature, e is quality of the currently best solution and e' is the quality of the perturbed solution. Because the performance of each instance of control software is stochastic, e and e' will be computed as the mean of a sample of 10 runs of the respective instance of control software. The temperature length determines the number of steps before the temperature cools down again. We set the value to 1, so that the cooling happens in every step. The cooling scheme that is then applied is the geometric cooling [23]. In geometric cooling, the updated temperature is computed as $T * \alpha$, where T is the current temperature and α is the cooling coefficient, which we set as $\alpha = 0.9782$. Additionally, the temperature resets to the initial value every 5000 simulations.

The source code of our implementation of IcePop is available at: https://github.com/keua/design-of-robot-swarms-by-optimization

2.3 Missions

All experiments were conducted with 20 robots on two missions AGGREGATION WITH AMBIENT CUES (AAC) and FORAGING.

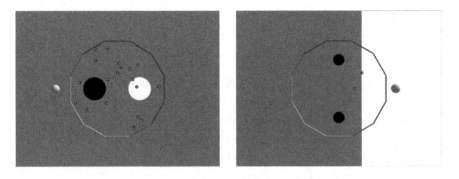

Fig. 1. The two missions: AAC (*left*) and FORAGING (*right*).

AAC. The arena contains two circles, one black, one white. A light source is placed on the side of the arena that contains the black circle (Fig. 1, left). The robots are tasked to aggregate on the black spot. The objective function $F_{\text{AAC}} = \sum_{t=0}^{T} N_t$ where N_t is the number of robots on the black circle at time step N_t.

Foraging. The arena contains two source areas in the form of black circles and a nest, as a white area. A light source is placed behind the nest to help the robots to navigate (Fig. 1, right). As the robots have no gripping capabilities, we consider an idealized version of foraging, where a robot is deemed to retrieve an object when it enters a source and then the nest. The goal of the swarm is to retrieve as many objects as possible. The objective function is $F_{\text{F}} = N_i$, where N_i is the number of retrieved objects.

2.4 Protocol

As each design process is stochastic, we run 20 independent designs for each design method, resulting in 20 instances of control software. The so obtained instances are then each assessed 10 times in the design context (what we call simulation) and another 10 times in a different simulation setting (what we call pseudo-reality). Pseudo-reality is a concept to evaluate the transferability of control software [25]. Instead of assessing the performance directly in reality, a different simulation context is used. Research has shown that control software that transfers well into reality also transfers well into pseudo-reality, while control software that transfers badly into reality also transfers badly into pseudo-reality.

The results are presented in notched box-and-whisker boxplots, giving a visual representation of the samples. In such a notched box-and-whisker box-plot, the horizontal thick line denotes the median of the sample. The lower and upper sides of the box are called upper and lower hinges and represent the 25th and 75th percentile of the observations, respectively. The upper whisker extends either up to the largest observation or up to 1.5 times the difference between upper hinge and median—whichever is smaller. The lower whisker is defined

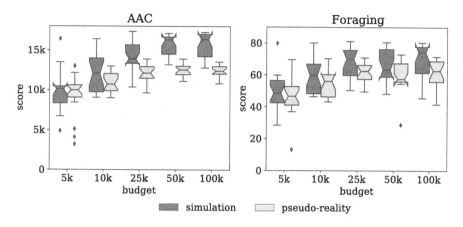

Fig. 2. Performance of control software created by `IcePop` for different budgets.

analogously. Small circles represent outliers (if any), that are observations that fall beyond the whiskers. Notches extend to $\pm 1.58 IQR/\sqrt{n}$, where IQR is the interquartile range and n = 20 is the number of observations. Notches indicate the 95% confidence interval on the position of the median. If the notches of two boxes do not overlap, the observed difference between the respective medians is significant [7].

3   Results

In this section we describe four experiments we conducted and the results we obtained. The instances of control software produced, the details of their performances, and videos of their execution on the robots are available as online supplementary material [24]. We also discuss possible reasons for the results.

3.1   Influence of the Budget

We conduct one experiment to investigate the influence of the budget on the performance of the generated control software. Designs with a smaller budget need less time to finish but usually produce results that perform less well in simulation. The higher the time the better usually the performance in simulation, but an overdesigning effect might be observed, where the improvement in simulation does not carry over to reality. We tested `IcePop` with five different budgets (5000, 10000, 25000, 50000 and 100000 simulations respectively).

The results displayed in Fig. 2 show the influence of the budget on the performance of the control software generated by `IcePop`. One trend that is apparent from the data, is that, as expected, a larger design budgets leads to control software that performs better in simulation. However the relative improvement diminishes and the performance seems to reach a peak around a budget of 50000 simulations.

Furthermore the performance in pseudo-reality improves initially with an increased budget. Here, however, the performance levels after the budget of 25000 simulations is reached and does not improve any further. This could be an indicator that the design reached the peak performance that is still transferable. Further designs might improve the performance in simulation but the transferability will suffer in return.

3.2   Influence of the Sample Size

We chose the Metropolis condition as the acceptance criterion in the component-based simulated annealing for `IcePop`. In its original definition it was defined to compare two single performance measures. As the evaluation of the performance of an instance of control software is stochastic, we sample several simulation runs. The mean of this sample is then supplied to the Metropolis condition.

In a second experiment, we investigate the influence of the sample size on the performance of the generated control software. Smaller sample sizes use less of the budget to evaluate one solution, allowing more solution candidates to be investigated. On the other hand, outliers will have a greater impact on the mean of the samples and thus the perceived performance. Larger sample sizes lead to the inverse effect. Fewer total solution candidates would be investigated but the performance of each individual solution candidate is more robust to outliers. We study the influence of the sample size on the performance of the generated control software by evaluating the performance in simulation and in pseudo-reality for three sample sizes: 5, 10, and 15. Additionally we test every variant on the three budgets that showed peak performance in the previous experiment (25000, 50000, and 100000 simulations).

Figure 3 shows the results for the three different variants of the sample size over the three investigated budgets. For a budget of 25000 simulations, all variants perform similar and no differences can be seen, both in simulation and pseudo-reality. In the case of a budget of 50000, the variant with a sample size of 10 samples performs slightly better than the other two variants, in the mission FORAGING when assessed in simulation. In pseudo-reality, this difference however is not present anymore. It could therefore very well be that this is simply a statistical artifact of the stochastic design process. For 100000 simulation runs, the three variants achieve a comparable performance again and only minor differences can be observed. All in all, the three different sample sizes that we compared show no noticeable differences.

3.3   Influence of the Restarting Mechanism

We conduct a third experiment, to investigate the influence of the restarting mechanism. Restarting resets the temperature to a higher value, allowing the design process to make bigger movements in the search space again. We investigate four different restarting mechanisms: fixed length (restarts after a fixed number of simulations, in this case every 5000 simulations), no restart (the temperature cools over the whole design process and is never restarted), reheat (the

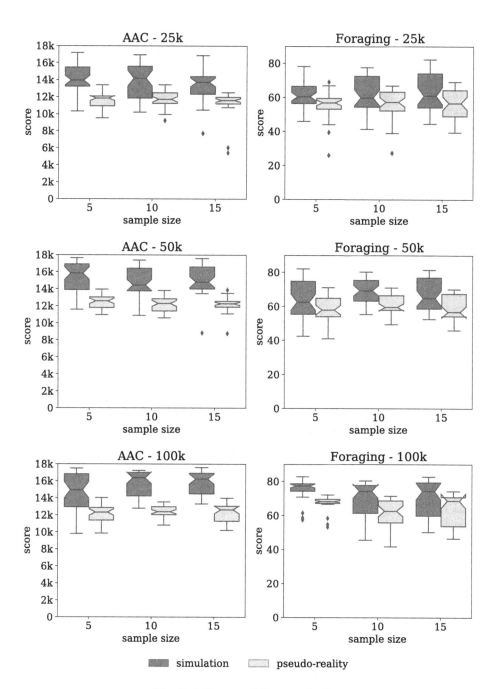

Fig. 3. Influence of the sample size.

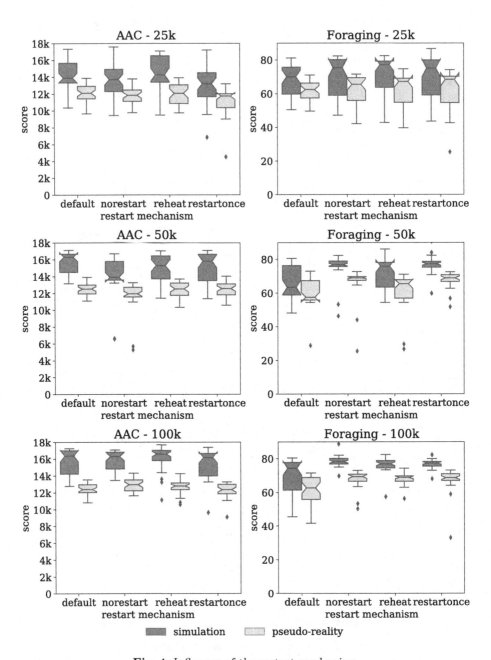

Fig. 4. Influence of the restart mechanism.

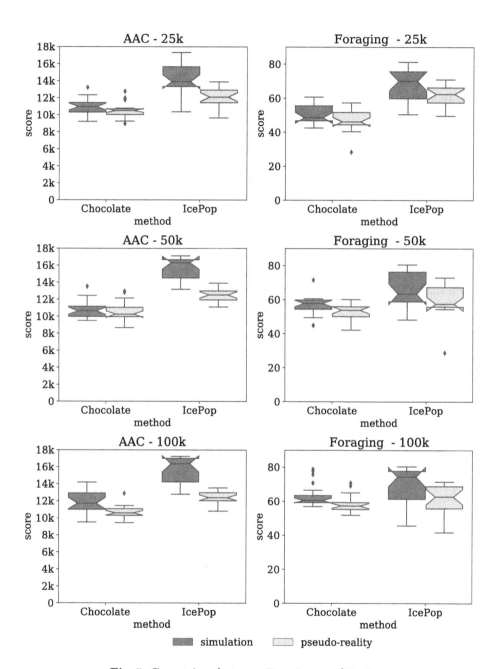

Fig. 5. Comparison between Chocolate and IcePop.

temperature is reset every 5000 simulations, the new temperature is set to the one that generated the biggest improvement so far), restart once (after the half of the budget is exhausted the temperature resets). We test all restarting mechanisms on budgets of 25000, 50000 and 10000 simulations.

Figure 4 shows the results for the different restarting mechanisms. The results for a budget of 25000 simulation runs show no difference between the four variants. In case of a budget of 50000 simulation runs all variants perform similarly in the mission AAC. In the mission FORAGING, the restarting mechanism that restarts every 5000 simulation runs performs worse than the other three variants. For a budget of 100000 simulation runs, all four variants perform similarly again. In the mission FORAGING, however, the fixed length restarting mechanism (default) shows a larger distribution than the other three variants.

In conclusion, the four different variants fail to produce noticeable differences in the performance of the generated control software.

3.4   Comparison with Chocolate

In the last experiment, we compare the performance of IcePop with Chocolate across three different budgets (25000, 50000 and 100000 simulations).

Figure 5 shows the comparison results of IcePop with Chocolate for budgets of 25000, 50000, and 100000 simulations respectively. Throughout all three budgets, it is apparent that IcePop performs better in simulation than Chocolate in both missions. In the mission AAC, the difference in performance is statistically significant.

Unfortunately the drop of performance when assessed in pseudo-reality is slightly larger for IcePop than for Chocolate. This could indicate that IcePop might be less transferable to real robots than Chocolate. Despite the larger performance drop, IcePop still performs better in pseudo-reality, and in AAC this difference in performance is also statistically significant.

Additionally, we have taken the best performing instance of control software of IcePop and Chocolate (with a design budget of 100k simulations) for each mission and directly applied it to a swarm of twenty real e-pucks. Videos of the performance of the control software on real robots can be found online [24].

4   Conclusions

In this work we have investigated a default configuration for simulated annealing in the context of automatic modular design. The results indicate that simulated annealing can be a viable candidate for the automatic modular design of robot swarms. Additionally, we have investigated the influence of some obvious variations to the simulated annealing on the performance of the automatic modular design. The component-based simulated annealing approach allowed us to easily implement these variants.

Simulated annealing is a well studied optimization algorithm with many proposed extensions, improvements and variants. A next step could be finding a suitable configuration of components that satisfies best the demands of the automatic modular design. Also, it would be interesting to apply IcePop to a broader range of missions.

Acknowledgements. The project has received funding from the European Research Council (ERC) under the European Union's Horizon 2020 research and innovation programme (grant agreement No 681872). Jonas Kuckling and Mauro Birattari acknowledge support from the Belgian *Fonds de la Recherche Scientifique* – FNRS.

References

1. Aarts, E., Korst, J., Michiels, W.: Simulated annealing. In: Burke, E.K., Kendall, G. (eds.) Search Methodologies: Introductory Tutorials in Optimization and Decision Support Techniques, pp. 187–210. Springer, Boston (2005). https://doi.org/10.1007/0-387-28356-0_7
2. Birattari, M., Stützle, T., Paquete, L., Varrentrapp, K.: A racing algorithm for configuring metaheuristics. In: Langdon, W.B., et al. (eds.) GECCO 2002: Proceedings of the Genetic and Evolutionary Computation Conference, pp. 11–18. Morgan Kaufmann Publishers, San Francisco (2002)
3. Birattari, M., Yuan, Z., Balaprakash, P., Stützle, T.: F-Race and iterated F-Race: an overview. In: Bartz-Beielstein, T., Chiarandini, M., Paquete, L., Preuss, M. (eds.) Experimental Methods for the Analysis of Optimization Algorithms, pp. 311–336. Springer, Heidelberg (2010). https://doi.org/10.1007/978-3-642-02538-9_13
4. Brambilla, M., Ferrante, E., Birattari, M., Dorigo, M.: Swarm robotics: a review from the swarm engineering perspective. Swarm Intell. **7**(1), 1–41 (2013). https://doi.org/10.1007/s11721-012-0075-2
5. Bredeche, N., Haasdijk, E., Prieto, A.: Embodied evolution in collective robotics: a review. Front. Robot. AI **5**, 12 (2018). https://doi.org/10.3389/frobt.2018.00012
6. Burke, E.K., Bykov, Y.: The late acceptance hill-climbing heuristic. Eur. J. Oper. Res. **258**(1), 70–78 (2017). https://doi.org/10.1016/j.ejor.2016.07.012
7. Chambers, J.M., Cleveland, W.S., Kleiner, B., Tukey, P.A.: Graphical Methods For Data Analysis. CRC Press, Belmont (1983)
8. Christensen, A.L., Dorigo, M.: Evolving an integrated phototaxis and hole-avoidance behavior for a swarm-bot. In: Rocha, L.M., Yaeger, L.S., Bedau, M.A., Floreano, D., Goldstone, R.L., Vespignani, A. (eds.) Artificial Life - ALIFE, pp. 248–254. MIT Press, Cambridge (2006). A Bradford Book
9. Dietterich, T.G., Kong, E.B.: Machine learning bias, statistical bias, and statistical variance of decision tree algorithms. Technical report, Department of Computer Science, Oregon State University, Corvallis, OR, USA (1995)
10. Dorigo, M., Birattari, M., Brambilla, M.: Swarm robotics. Scholarpedia **9**(1), 1463 (2014). https://doi.org/10.4249/scholarpedia.1463
11. Ferrante, E., Turgut, A.E., Duéñez-Guzmán, E.A., Dorigo, M., Wenseleers, T.: Evolution of self-organized task specialization in robot swarms. PLoS Comput. Biol. **11**(8), e1004273 (2015). https://doi.org/10.1371/journal.pcbi.1004273
12. Francesca, G., Birattari, M.: Automatic design of robot swarms: achievements and challenges. Front. Robot. AI **3**(29), 1–9 (2016). https://doi.org/10.3389/frobt.2016.00029

13. Francesca, G., et al.: AutoMoDe-Chocolate: automatic design of control software for robot swarms. Swarm Intell. **9**(2–3), 125–152 (2015). https://doi.org/10.1007/s11721-015-0107-9
14. Francesca, G., Brambilla, M., Brutschy, A., Trianni, V., Birattari, M.: AutoMoDe: a novel approach to the automatic design of control software for robot swarms. Swarm Intell. **8**(2), 89–112 (2014). https://doi.org/10.1007/s11721-014-0092-4
15. Franzin, A., Stützle, T.: Revisiting simulated annealing: a component-based analysis. Comput. Oper. Res. **104**, 191–206 (2019). https://doi.org/10.1016/j.cor.2018.12.015
16. Garattoni, L., Francesca, G., Brutschy, A., Pinciroli, C., Birattari, M.: Software infrastructure for e-puck (and TAM). Technical report TR/IRIDIA/2015-004, IRIDIA, Université libre de Bruxelles, Belgium (2015)
17. Geman, S., Bienenstock, E., Doursat, R.: Neural networks and the bias/variance dilemma. Neural Comput. **4**(1), 1–58 (1992). https://doi.org/10.1162/neco.1992.4.1.1
18. Gutiérrez, Á., Campo, A., Dorigo, M., Donate, J., Monasterio-Huelin, F., Magdalena, L.: Open e-puck range & bearing miniaturized board for local communication in swarm robotics. In: Kosuge, K. (ed.) IEEE International Conference on Robotics and Automation, ICRA, pp. 3111–3116. IEEE, Piscataway (2009). https://doi.org/10.1109/ROBOT.2009.5152456
19. Hajek, B.: Cooling schedules for optimal annealing. Math. Oper. Res. **13**(2), 311–329 (1988). https://doi.org/10.1287/moor.13.2.311
20. Hasselmann, K., et al.: Reference models for AutoMoDe. Technical report TR/IRIDIA/2018-002, IRIDIA, Université libre de Bruxelles, Belgium (2018)
21. Hauert, S., Zufferey, J.C., Floreano, D.: Evolved swarming without positioning information: an application in aerial communication relay. Auton. Robots **26**(1), 21–32 (2009). https://doi.org/10.1007/s10514-008-9104-9
22. Hoos, H., Stützle, T.: Stochastic Local Search: Foundations & Applications, 1st edn. Morgan Kaufmann Publishers, San Francisco (2005). https://doi.org/10.1016/B978-1-55860-872-6.X5016-1
23. Kirkpatrick, S., Gelatt Jr., C.D., Vecchi, M.P.: Optimization by simulated annealing. Science **220**(4598), 671–680 (1983). https://doi.org/10.1126/science.220.4598.671
24. Kuckling, J., Ubeda Arriaza, K., Birattari, M.: AutoMoDe-IcePop: automatic modular design of control software for robot swarms using simulated annealing (2020). Supplementary material. http://iridia.ulb.ac.be/supp/IridiaSupp2020-003/
25. Ligot, A., Birattari, M.: Simulation-only experiments to mimic the effects of the reality gap in the automatic design of robot swarms. Swarm Intell. **14**(1), 1–24 (2019). https://doi.org/10.1007/s11721-019-00175-w
26. Lundy, M., Alistair, M.: Convergence of an annealing algorithm. Math. Program. **34**(1), 111–124 (1986). https://doi.org/10.1007/BF01582166
27. López-Ibáñez, M., Dubois-Lacoste, J., Pérez Cáceres, L., Birattari, M., Stützle, T.: The irace package: iterated racing for automatic algorithm configuration. Oper. Res. Perspect. **3**, 43–58 (2016). https://doi.org/10.1016/j.orp.2016.09.002
28. Metropolis, N., Rosenbluth, A.W., Rosenbluth, M.N., Teller, A.H., Teller, E.: Equation of state calculations by fast computing machines. J. Chem. Phys. **21**(6), 1087–1092 (1953). https://doi.org/10.1063/1.1699114

29. Mitra, D., Romeo, F., Sangiovanni-Vincentelli, A.: Convergence and finite-time behavior of simulated annealing. In: 1985 24th IEEE Conference on Decision and Control, pp. 761–767. IEEE Press, Piscataway (1985). https://doi.org/10.1109/CDC.1985.268600

30. Mondada, F., et al.: The e-puck, a robot designed for education in engineering. In: Gonçalves, P., Torres, P., Alves, C. (eds.) Proceedings of the 9th Conference on Autonomous Robot Systems and Competitions, pp. 59–65. Instituto Politécnico de Castelo Branco, Castelo Branco (2009)

31. Nedjah, N., Silva Junior, L.: Review of methodologies and tasks in swarm robotics towards standardization. Swarm Evol. Comput. **50**, 100565 (2019). https://doi.org/10.1016/j.swevo.2019.100565

32. Nikolaev, A.G., Jacobson, S.H.: Simulated annealing. In: Gendreau, M., Potvin, J.Y. (eds.) Handbook of Metaheuristics. International Series in Operations Research & Management Science, vol. 146, pp. 1–39. Springer, Boston (2010). https://doi.org/10.1007/978-1-4419-1665-5_1

33. Quinn, M., Smith, L., Mayley, G., Husbands, P.: Evolving controllers for a homogeneous system of physical robots: structured cooperation with minimal sensors. Philos. Trans. R. Soc. Lond. Ser. A Math. Phys. Eng. Sci. **361**(1811), 2321–2343 (2003). https://doi.org/10.1098/rsta.2003.1258

34. Silva, F., Duarte, M., Correia, L., Oliveira, S.M., Christensen, A.L.: Open issues in evolutionary robotics. Evol. Comput. **24**(2), 205–236 (2016). https://doi.org/10.1162/EVCO_a_00172

35. Trianni, V., López-Ibáñez, M.: Advantages of task-specific multi-objective optimisation in evolutionary robotics. PLoS One **10**(8), e0136406 (2015). https://doi.org/10.1371/journal.pone.0136406

36. Trianni, V., Nolfi, S.: Self-organizing sync in a robotic swarm: a dynamical system view. IEEE Trans. Evol. Comput. **13**(4), 722–741 (2009). https://doi.org/10.1109/TEVC.2009.2015577

37. Yang, G.Z., et al.: The grand challenges of Science Robotics. Sci. Robot. **3**(14), eaar7650 (2018). https://doi.org/10.1126/scirobotics.aar7650

Evaluation of Alternative Exploration Schemes in the Automatic Modular Design of Robot Swarms

Gaëtan Spaey, Miquel Kegeleirs⑩, David Garzón Ramos⑩,
and Mauro Birattari[(✉)] ⑩

IRIDIA, Université Libre de Bruxelles, Brussels, Belgium
mbiro@ulb.ac.be

Abstract. The swarm robotics literature has shown that complex tasks
can be solved by large groups of simple robots interacting with each other
and their environment. Most of these tasks require the robots to explore
their environment, making exploration a building block of the behaviors
of robot swarms. However, exploration schemes have rarely been thor-
oughly evaluated, especially in the context of automatic design. This is
the case with AutoMoDe, an automatic modular design approach that
designs control software by assembling predefined mission-agnostic modules that embed fixed and arbitrarily selected exploration schemes. In
this paper, we study the influence of different exploration schemes on the
automatic design of robot swarms. To do so, we introduce AutoMoDe-
Coconut, a new variant of AutoMoDe with multiple configurable exploration schemes embedded within its modules. We test Coconut both in
bounded and unbounded workspaces and we compare the results with
those of AutoMoDe-Chocolate in order to understand the impact of the
new exploration schemes. The results show that Coconut is prone to
select exploration schemes that fulfill the requirements of the mission in
hand. However, Coconut does not perform better than Chocolate, even in
situations where the only exploration schemes available to Chocolate are
at an apparent disadvantage. We conjecture that the overall exploration
capabilities of the swarm are not the mere reflection of individual-level
exploration schemes but result from a more complex interaction between
the atomic behaviors of the individuals.

Keywords: Automatic design · Exploration · Random walk

1 Introduction

A robot swarm is a large group of robots whose collective behavior results
from local interactions of the robots between themselves and with their

GS and MK contributed equally to this work and should be considered as co-first
authors. The software used in the experiments was implemented by GS and MK. The
experiments were designed by DGR and performed by GS and MK. The paper was
drafted by GS and MK and edited by MK and MB; all authors read and commented
the final version. The research was directed by MB.

B. Bogaerts et al. (Eds.): BNAIC 2019/BENELEARN 2019, CCIS 1196, pp. 18–33, 2020.
https://doi.org/10.1007/978-3-030-65154-1_2

environment [9]. A robot swarm operates without relying on any external structure or any form of centralized control [1, 32]. These characteristics make swarms of robots scalable, robust and flexible.

Unfortunately, the design of control software for robot swarms is a complex activity. Indeed, there is no reliable way to anticipate the global behavior of a swarm of robots based on the behavior of a single individual [12]. It is therefore common to resort to automatic design, for which multiple methods have been developed [10, 26, 37]. In particular, AutoMoDe [14] is an automatic modular design approach that produces control software by assembling preexisting software modules into an appropriate modular architecture'e.g., a finite-state machine or a behavior tree. The possible states come from a finite set of atomic behaviors, such as the attraction to light sources or the repulsion from other robots. A few methods belonging to the AutoMoDe family have been proposed so far. Most of them are variants of Vanilla, the first method proposed that meet the specifications of AutoMoDe [14]. In these variants, most of the atomic behaviors embed the same exploration scheme: ballistic motion.

We foresee that other exploration schemes, such as random walks [11, 27, 31, 38], could improve the exploration capabilities of robot swarms automatically generated via AutoMoDe.

To study the influence of different exploration schemes in automatic modular design of robot swarms, we introduce AutoMoDe-Coconut, a new variant of AutoMoDe able to select different exploration schemes. Following the tenets of the automatic offline design of robot swarms [4], we assess the capabilities of Coconut to design control software for missions that require the robot swarm to explore in different manners. To this aim, we conduct experiments on two classes of missions using realistic simulations and real robots experiments. We compare the performance of Coconut against the one of the state-of-the-art modular design method AutoMoDe-Chocolate [13]. We expect Coconut to outperform Chocolate in at least one class of mission thanks to its extended exploration capabilities. To the best of our knowledge, this is the first time different exploration schemes are compared in the context of automatic design.

The paper is structured as follows. In Sect. 2, we discuss related work in automatic design and exploration. In Sect. 3, we present Coconut, the automatic modular design method we investigate in the paper. In Sect. 4 we describe the experimental setup. In Sect. 5, we illustrate the results of the experiments. In Sect. 6, we conclude the paper and we sketch future research.

2   Related Work

In single-robot systems, the control software is typically designed by hand by a human developer as the behavior of the robot is easy to derive from its specifications. In swarm robotics however, the link between the behavior of the individual robots that one should program and the global behavior of the swarm that one wishes to obtain is often particularly complex. Indeed, it is difficult to anticipate the behavior of a swarm solely based on the individual behavior of the robots [6]. The control software of the individual robots is therefore a trial and error process, which is time consuming, prone to bias and errors, and difficult to replicate [5]. Automatic design appears to be a promising way to overcome the difficulties of generating control software for robot swarms [12].

Neuro-evolutionary robotics is the classical automatic design approach adopted in swarm robotics [8,26]. In this approach, robots are controlled by a neural network whose parameters (and possibly the structure) are optimized using an evolutionary algorithm in an off-line process based on computer simulations [29,35,36]. The inputs of the neural networks are the readings of the sensors and outputs are the commands to be fed to the actuators. Unfortunately, the neuro-evolutionary approach is known to produce control software that crosses the reality gap poorly [19,33]. Indeed, a noticeable drop in performance can be often observed when neural networks optimized in simulation are tested on real robots. This is the result of a sort of *overfitting* of the control software to the simulator, which prevents it to then generalize to the real world [22].

An alternative approach to automatic design is the automatic modular design method proposed by Francesca et al. AutoMoDe [14]. The original idea behind AutoMoDe is to inject a bias in the automatic design process by increasing the granularity of the control software architecture. This reduces the risk of overfitting the simulator and eventually increases the chance that the control software produced crosses the reality gap successfully, generalizing properly to reality. Multiple variants of AutoMoDe have been developed so far [13,18,21].

However, the exploration scheme used by AutoMoDe, ballistic motion, was selected arbitrarily from Vanilla and has been kept in all the following studies without a further discussion. Recent works have shown the relevance of the exploration scheme in robot swarms. Common random walks (Brownian motion [11], correlated random walk [31], Lévy walk [38] and Lévy taxis [27]) have been evaluated by Dimidov et al. [7] with a swarm of Kilobots. An optimal parametrization for these random walks was found with this configuration. Kegeleirs et al. [20] evaluated the same random walks, along with ballistic motion, for mapping with ten e-pucks and found that the parametrization does not generalize to other robotic platforms. Similarly, Ramachandran et al. [30] performed distributed mapping with three robots using a custom variant of Lévy walk. To the best of our knowledge, no study has been published that evaluates the performance of the aforementioned exploration schemes in the context of the automatic design of collective behaviors for robot swarms.

3   AutoMoDe-Coconut

Coconut builds on Chocolate [13]. As Chocolate, it belongs to the AutoMoDe class of methods originally defined by Francesca et al. [14]. These methods automatically generate control software by assembling predefined, mission-agnostic software modules. Like Chocolate, Coconut produces control software for the e-puck platform [25]. The control software produced by Coconut, like the one produced by Chocolate, has the form of a probabilistic finite-state machine.

The modules are either low-level behaviors to be used as states of the state machine or conditions to be associated with its edges. Conditions determine whether a transition should happen or not. Modules may have tunable parameters that modify their functioning. The topology of the probabilistic finite-state machine, the behaviors and the conditions to be included, and the value of their parameters are determined by an optimization algorithm that maximizes a mission-specific performance measure. The optimization algorithm adopted in Coconut, as well as in Chocolate, is Iterated F-race (irace) [23]. The only difference between Coconut and Chocolate is that, in Coconut, the modules have a parameter controlling the type of exploration scheme to use, whereas in Chocolate the scheme is fixed for each module. Indeed, Coconut embeds three different exploration schemes within its modules: ballistic motion with random rotations, ballistic motion with vector field and, random walk. As Coconut is identical to Chocolate in all other aspects, the discussion on performance differences can focus on the sole influence of these exploration schemes.

3.1   Robot Platform

Coconut produces control software for the e-puck platform, extended with three hardware modules: the Overo Gumstix, the ground sensor, and the range and bearing. The e-puck is a circular two-wheeled robot, whose diameter is approximately 70 mm. It has 8 IR transceivers, positioned all around its body, that work both as light and proximity sensors. The Overo Gumstix module is a single-board computer that allows the e-puck to run Linux. The ground sensor module allows the e-puck to perceive the color of the floor. The range-and-bearing module [16] is an infrared communication device for local sensing and messaging. It operates by broadcasting a ping signal that can be received by robots within a range of about 0.7 m from the sender. A robot that receives a ping is able to estimate the relative position of the sender in polar coordinates. The capabilities of the e-puck platform are formally defined by the reference model RM 1.1 [17], see Table 1.

3.2   Set of Modules

Coconut's modules are built upon those of Chocolate. They comprise 6 behaviors and 6 transitions:

Behaviors:

- rambling[1]: the robot explores randomly its environment;
- stop: the robot stands still;
- phototaxis: the robot goes towards the light source, if perceived;
- anti-phototaxis: the robot goes away from the light, if perceived;
- attraction: the robot goes towards its neighboring peers, if perceived;
- repulsion: the robot goes away from its neighboring peers, if perceived.

---

[1] Originally, this module was called exploration [14]. In this paper, we changed its denomination to avoid confusion with the notion of exploration scheme.

Table 1. Reference model RM1 of the e-puck robot [17]. RM1 abstracts sensors and actuators by defining the input and the output variables that are made available to the control software at each control step. Sensors are defined as input variables: the control software can only read them. Actuators are defined as output variables: the control software can only write them. Input and output variables are updated with a period of 100 ms.

Sensor/Actuator	Variables
Proximity	$prox_i \in [0, 1]$, with $i \in \{1, 2, ..., 8\}$
Light	$light_i \in [0, 1]$, with $i \in \{1, 2, ..., 8\}$
Ground	$ground_i \in \{white, gray, black\}$, with $i \in \{1, 2, 3\}$
Range-and-bearing	$n \in [0, 20]$ $r_m \in [0, 0.70]$, with $m \in \{1, 2, ..., 20\}$ $b_m \in [0, 2\pi]$ rad, with $m \in \{1, 2, ..., 20\}$
Wheels	v_l, $v_r \in [-0.12, 0.12]$ m/s

Conditions:

- black-floor: change state if floor is black;
- white-floor: change if it is white;
- gray-floor: change if it is gray;
- neighbor-count: change if sufficiently many neighboring peers are perceived;
- inverted-neighbor-count: change if they are sufficiently few;
- fixed-probability: change state with a fixed probability.

The behaviors are identical to those of Chocolate except for the exploration scheme adopted. In Chocolate, fixed default exploration schemes are used in the rambling module (ballistic motion with random rotations) as well as in the phototaxis, anti-phototaxis, attraction and repulsion modules when no light/no neighboring peers are perceived (ballistic motion with vector field). In Coconut, these five modules do not use a default exploration scheme but have instead a new ϵ parameter that has 3 possible values: BMVF, BMRR, and RW.

If $\epsilon = $ **BMVF**, the exploration scheme is a ballistic motion with vector field. The robot follows the two-dimensional vector $w = w_b - w_o$, where w_b represents the ballistic vector and w_o the perceived obstacle vector. The ballistic vector is trivially defined as $w_b = (1, \angle 0)$ and represent a straight motion. The obstacle vector w_o is calculated with Eq. 1.

$$w_o = \sum_{i=1}^{8} (r_i, \angle b_i) \tag{1}$$

Where r_i represents the reading of i, one of the eight proximity sensors of the robot and b_i the angle between this sensor and the front of the robot. The vector w_o therefore represents the average position of the sensed obstacle(s) as the sum of the eight vectors corresponding to proximity readings.

Table 2. Different value ranges of the key parameters of the random walks, along with the corresponding distributions [7,27].

	Move length	μ	Turning angle	ρ
Brownian motion	Asymptotically Gaussian-like	3	Uniform	0
Correlated random walk	Asymptotically Gaussian-like	3	Wrapped Cauchy	$\in [0,1]$
Lévy walk	Power law	$\in]1,3]$	Uniform	0
Lévy taxis	Power law	$\in]1,3]$	Wrapped Cauchy	$\in [0,1]$

If $\epsilon = $ **BMRR**, the exploration scheme is a ballistic motion with random rotations. The robot moves in a straight line. When it encounters an obstacle, it turns on itself for a random number of control cycles uniformly chosen between $[0, \tau]$, where τ is a parameter of the module. The parameter τ is an integer in the range $[0, 100]$.

If $\epsilon = $ **RW**, the exploration scheme is a random walk. The robot follows the two-dimensional vector $w = w_{Lt} - w_o$, where w_{Lt} represents the Lévy taxis vector and w_o is calculated with Eq. 1, as in the ballistic motion with vector field. The Lévy taxis vector is calculated as $w_{Lt} = (1, \angle T_a)$ where T_a is the turning angle defined by Eq. 2.

$$T_a = 2 \arctan \left(\frac{1 - \rho}{1 + \rho} \tan \left(\pi (r - 0.5) \right) \right) + bias \tag{2}$$

The turning angle changes after a number of control cycles governed by the movement length M_l defined by Eq. 3.

$$M_l = L_{min} r^{\frac{1}{1-\mu}} \tag{3}$$

These equations depend on the parameters μ and ρ, 2 parameters of the module. The parameter μ is real-valued and chosen in the range $]1, 3]$. The parameter ρ is real-valued as well and chosen in the range $[0, 1]$. Table 2 presents the values of μ and ρ for the major state-of-the-art random walks that can be modeled by Eqs. 2 and 3. These additional parameters have also an effect on the search space size: it is larger for Coconut than for Chocolate.

3.3   Automatic Design Process

Coconut produces control software in the form of probabilistic finite-state machines. The topology of the probabilistic finite-state machine, the modules to be included and their parameters are defined by an optimization process. The space of the probabilistic finite-state machines that Coconut can possibly generate is constrained to those comprising at most 4 states having each at most 4 outgoing edges.

As an optimization algorithm, Coconut uses the implementation of Iterated F-race provided by the R package irace [23] with its default parameters. Iterated

F-race is based on F-race [2], a racing procedure where a set of candidate solutions are randomly sampled and then sequentially evaluated, over a set of test cases, to eventually select the most suitable one. Along the sequential evaluation of candidate solutions, a Friedman test is repeatedly performed to identify candidate solutions that perform significantly worse than at least another one. These solutions are discarded so that the evaluation can focus on the best ones. The algorithm terminates when only one candidate solution remains or when a predefined budget of evaluations is depleted. Iterated F-race consists of multiple iterations of F-race. After the first iteration, each subsequent one operates on a set of candidate solutions that are sampled around those that the previous iteration selected as the best ones. The algorithm terminates when a predefined budget of evaluations is depleted.

Within the optimization process, simulations are performed using the ARGoS3 simulator [28], version beta 48, together with the argos3-epuck library [15]. ARGoS3 is a modular multi-physics robot simulator specifically conceived to simulate robot swarms. Coconut uses the 2D dynamic physics engine of ARGoS3 to simulate the robots and the environment. The argos3-epuck library provides low-level implementations of the sensors and actuators of the e-puck robot with fine control on noise levels for all actuators and sensors. ARGoS3 and the argos3-epuck library inject a realistic level of sensor and actuator noise in all simulations as suggested by Miglino et al. [24] as a good practice for reducing the impact of the reality gap.

4   Experimental Setup

In order to assess the performance impact of the new exploration schemes integrated in its modules, we compare Coconut to Chocolate on a set of missions, both in bounded and unbounded workspaces. Similarly to previous studies in automatic modular design, we also compare Chocolate and Coconut to Evostick, an automatic design method that implements a typical evolutionary robotics setup. Evostick was introduced in Francesca et al. [14] to define a yardstick against which AutoMoDe variants can be compared. We expect Coconut to produce results similar to those of Chocolate in bounded workspace but to outperform Chocolate in unbounded workspace. The choice of the missions is motivated by the need to challenge both the general problem-solving capabilities of the two methods and their exploration capabilities. Therefore, the missions consist in missions already used to test other AutoMoDe variants as well as a new mission specifically targeting the exploration capabilities offered by the new exploration schemes. The chosen missions are aggregation, foraging, and grid exploration.

4.1   Missions

Each mission takes place in a dodecagonal workspace of $4.91\,\mathrm{m}^2$ surrounded by walls that are tall enough to prevent the robots from seeing anything beyond them. The floor is gray with the exception of black or white areas specific to each mission. All missions are performed by a swarm of 20 e-puck robots for a duration of 120 s. In the following descriptions, the coordinates are in meters with the origin of the axes at the center of the workspace. The x axis points right and the y axis points up. The three missions are detailed below.

Aggregation: The robots must aggregate as fast as possible on a black spot at the center of the workspace. The floor is completely gray except for a black circular area of diameter 0.60 m at the center of the workspace. At the beginning of the experiment, the 20 robots are randomly placed in the whole workspace. Figure 1a shows the workspace of the mission. The performance of the swarm is measured by the sum of the time spent, in seconds, by each robot in the black area during the whole duration of the mission. Formally:

$$F_{aggregation} = \sum_{i=1}^{N} T_i \qquad (4)$$

Where $N = 20$ is the number of robots and T_i is the aggregated time spent in the black area by robot i during the whole duration of the mission.

Foraging: The robots must retrieve as many objects as possible from two sources and drop them in a specific area, the nest. The sources and nest are represented respectively by two black spots and a white area. The two black spots are black circular areas of diameter 0.30 m located at the coordinates $(0, 0.75)$ and $(0, -0.75)$. The white area covers the whole region of the workspace with $x > 0.60$. Moreover, a light source is placed behind the nest at coordinates $(1.25, 0)$ at 0.75 m from the ground. Figure 1b shows the workspace of the mission. Since the e-puck robot doesn't have grasping capabilities, the transportation of objects is abstracted. Therefore, it is supposed that a robot grabs an object (if it isn't already holding one) when it enters a source and drops the object (if it has one) when it enters the nest. At the beginning of the experiment, the 20 robots are randomly placed in the workspace. The performance of the swarm is measured by the sum of the number of objects retrieved by each robot, during the whole duration of the mission. Formally:

$$F_{foraging} = \sum_{i=1}^{N} O_i \qquad (5)$$

Where $N = 20$ is the number of robots and O_i is the number of objects retrieved by the robot i.

Grid Exploration: The robots must explore and cover as much space as possible. The floor is completely gray. At the beginning of the experiment, the 20 robots are randomly placed in the workspace. Figure 1c shows the workspace of the mission. In order to measure the performance of the swarm, the arena is divided in a grid of 10 tiles by 10 tiles. For each tile, we retain the time t elapsed since the last time it was visited by a robot. Each time the tile is visited by a robot, this time is reset to 0. The performance of the swarm is measured by the sum over all control cycles of the opposite of the average time t over all the tiles.

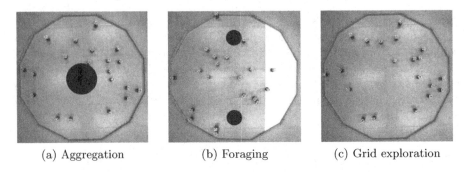

(a) Aggregation   (b) Foraging   (c) Grid exploration

Fig. 1. Workspaces of the 3 bounded missions, including an example of initial positions for the robots

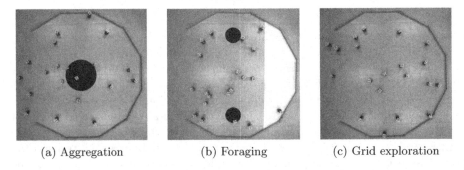

(a) Aggregation   (b) Foraging   (c) Grid exploration

Fig. 2. Workspaces of the 3 unbounded missions, including an example of initial positions for the robots

Formally:

$$F_{gridexploration} = \sum_{i=1}^{N_{cc}} \left(\frac{1}{N_{tiles}} \sum_{j=1}^{N_{tiles}} -t_{ij} \right) \tag{6}$$

Where N_{cc} is the number of control cycles for the whole experiment, N_{tiles} is the number of tiles and t_{ij} is the time, at the control cycle i, since the tile j was crossed by a robot.

For each of these missions, we evaluated an alternative version in an unbounded workspace. The only difference between a mission and its unbounded counterpart is that 3 walls have been removed from the workspace of the unbounded workspace mission. The 3 walls are the same for all the missions, namely the leftmost wall and its 2 neighbors. Figure 2 displays the workspaces of those alternative unbounded missions. These 2 sets of missions constitute the bounded and the unbounded classes of missions studied in this paper.

4.2   Protocol

Coconut, Chocolate and Evostick are executed 10 times on each of the 3 missions of each class with a budget of 100.000 evaluations. This design process produces 10 instances of control software per mission and per method. Each of these instances is then evaluated once on its respective mission[2]. The results of these evaluations are then presented mission by mission.

Then, each instance is uploaded on real e-pucks and evaluated once in a real environment with the same geometry and features as in the simulation. The results of these evaluations are also presented for each mission.

The evaluation of the performance on each mission is represented by notched box plots. For each mission, the score obtained in simulation and in reality for Coconut, Chocolate and Evostick is reported. Statements about the relative performance of the three methods on a specific mission are supported by the confidence intervals of those box plots. The evaluation of the aggregated performance over all of the missions is represented by a Friedman test. Once again, statements about the relative performance of the two methods are supported by the confidence intervals of this test. Any statement like "A performs *significantly* better/worse than B" means that the confidence intervals of the box plots of the scores obtained or the Friedman test for A and B do not overlap.

In order to interpret the observed performance of the automatic modular design methods, one needs to have some insight into the modules used by the two variants of AutoMoDe. We use two ways to measure the use of the different modules during a mission. The first one consists in counting, for each module, the proportion of instances of control software using this module in their finite-state machine. While this measurement gives some information about the finite-state machines and the behavior of the control software, it also shows modules that might not actually be used at runtime. Indeed, some states of the finite-state machines can be bypassed completely by high-probability transitions, making them useless. The second measurement is the average (across all of the robots of the swarm and all instances of control software of the mission) of the proportion of time each robot uses the behavior of each module. While this measure gives a better idea of the actual use of the different modules at runtime, it fails to differentiate important modules used for a short time and useless modules used as transitions. For that reason, the two measurements are compared.

5   Results

We present the qualitative analysis of the results. Demonstrative videos, code, and additional results are available in [34]. The performance of Chocolate, Coconut and Evostick on the two classes of missions are shown in Fig. 3. For all missions, Evostick performs the same or better than both AutoMoDe methods in simulation but completely fails in reality. This is coherent with previous results obtained in the literature [13,18]. For Chocolate and Coconut, the results in

---

[2] This protocol has been used in [13,14,18,21,22] and is further discussed in [3].

simulation are close to those in reality for all the missions and for both methods although a small reality gap can be seen. An analysis of the exploration schemes used by Coconut for the different missions is shown in Fig. 4. We observe that Coconut selects the ballistic motion for the bounded missions to promote exploration. Indeed, ballistic motion allows the robots to cover larger distances. For the unbounded missions, Coconut switches to random walk to promote exploitation. The random walk tends to keep robots in the same area and hence reduces the risks to lose robots. In this sense, the exploration scheme has an influence in the unbounded class of missions.

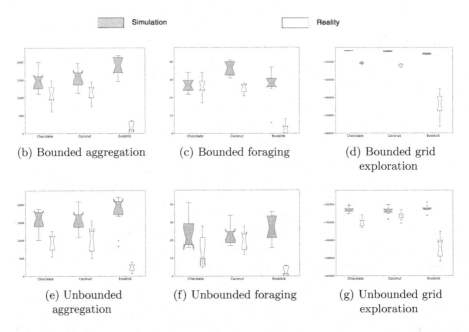

Fig. 3. Performance of Chocolate, Coconut and Evostick on aggregation, foraging and grid exploration, in bounded (top) and unbounded (bottom) workspaces. The higher the better.

Performance-wise, Coconut performs similarly to Chocolate in most missions. Differences between Chocolate and Coconut can only be observed for the bounded versions of foraging and grid exploration. However, these differences do not result from the exploration capabilities of Coconut but rather from the difference between the search space size of both methods. Indeed, Coconut has a larger search space and hence explores more solutions. Eventually, Coconut can find a solution that Chocolate cannot produce. In particular, this is the case for the bounded foraging mission. On the contrary, Chocolate will explore fewer solutions and converge to an optimal solution faster than Coconut. Eventually, this can translate into a slightly better performance, like for the bounded grid

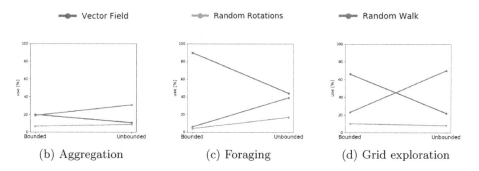

(b) Aggregation   (c) Foraging   (d) Grid exploration

Fig. 4. Runtime use of the ballistic motion with vector field, ballistic motion with random rotations and random walk exploration schemes in control software designed by Coconut for aggregation, foraging and grid exploration, in bounded and unbounded workspaces.

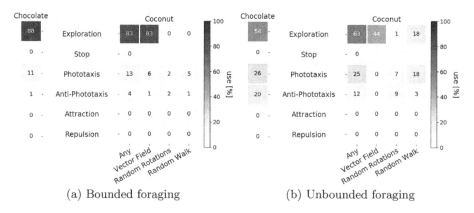

(a) Bounded foraging   (b) Unbounded foraging

Fig. 5. Runtime use of the modules in control software designed by Chocolate and Coconut for foraging in bounded (left) and unbounded (right) workspaces.

exploration. All in all, there is no improvement from the addition of new exploration schemes, even for the unbounded class of missions for which relying only on ballistic motion is an apparent disadvantage.

Therefore, we analyze the finite-state machines produced by Chocolate and Coconut on bounded and unbounded missions. We focus here on foraging but the following observations can also be made on the other missions. We can see in Fig. 5 that, in the unbounded mission, both Chocolate and Coconut decrease their use of the exploration module to rely more on the light (phototaxis and anti-phototaxis). The light is indeed at the opposite of the open part of the workspace and helps the robots to stay within the workspace. Considering that the performance of Chocolate and Coconut are similar, random walk does not help Coconut to achieve a better behavior. Therefore, we conjecture that Chocolate is able to adapt to different classes of missions by combining the modules at its disposal, without relying on different exploration schemes. In this sense, the exploration capabilities of Chocolate emerge from the interaction between its different modules rather than being the direct result of specific embedded exploration schemes.

6   Conclusions

We introduced Coconut, an automatic modular design method able to select different exploration schemes for each of its modules. We evaluated Coconut on three missions in bounded and unbounded workspaces. We observed that Coconut is prone to select exploration schemes that fit the requirements of the mission at hand. In bounded workspace, the control software produced uses mainly ballistic motion as it allows robots to cover bigger distances than random walks and promotes hence exploration. On the contrary, in unbounded workspaces, the control software produced uses mainly instances of random walk as it promotes exploitation behaviors that help maintaining the robots within the workspace. In this sense, the influence of the exploration scheme is only relevant for the class of missions in which the workspace is unbounded.

We also compared Coconut to Chocolate, the state-of-the-art automatic modular design method. Performance-wise, we could not observe a conclusive difference between the control software produced by Chocolate and Coconut, even in unbounded workspaces. Other exploration schemes do not improve the performance of the swarm as we expected but the results are still interesting as they allow us to make the following observations.

Even though Chocolate could only rely on ballistic motion as exploration scheme, it yielded a similar performance to Coconut in unbounded workspaces. Chocolate was hence able to design a control software preventing the robots to leave the workspace by combining its different modules. This means that exploration capabilities come from the interaction between atomic behaviors and not only from the exploration schemes embedded in the modules. In this sense, we saw that AutoMoDe adjusts to produce appropriate exploration strategies for the task at hand.

For the class of missions conceived so far, ballistic motion has proven to be a sufficiently appropriate exploration scheme. Still, whether random walk exploration could be a suitable solution in other contexts needs to be further explored.

Acknowledgements. The project has received funding from the European Research Council (ERC) under the European Union's Horizon 2020 research and innovation programme (grant agreement No 681872). Mauro Birattari acknowledges support from the Belgian Fonds de la Recherche Scientifique – FNRS. David Garzón Ramos acknowledges support from the Colombian Administrative Department of Science, Technology and Innovation – COLCIENCIAS.

References

1. Beni, G.: From swarm intelligence to swarm robotics. In: Şahin, E., Spears, W.M. (eds.) SR 2004. LNCS, vol. 3342, pp. 1–9. Springer, Heidelberg (2005). https://doi.org/10.1007/978-3-540-30552-1_1
2. Birattari, M., Stützle, T., Paquete, L., Varrentrapp, K.: A racing algorithm for configuring metaheuristics. In: Langdon, W., et al. (eds.) Proceedings of the Genetic and Evolutionary Computation Conference, GECCO, pp. 11–18. Morgan Kaufmann, San Francisco (2002)
3. Birattari, M.: Notes on the estimation of the expected performance of automatic methods for the design of control software for robot swarms. Technical report TR/IRIDIA/2020-010, IRIDIA, Université Libre de Bruxelles, Belgium (2020)
4. Birattari, M., Ligot, A., et al.: Automatic off-line design of robot swarms: a manifesto. Front. Robot. AI **1**(1), 1 (2019)
5. Bozhinoski, D., Birattari, M.: Designing control software for robot swarms: software engineering for the development of automatic design methods. In: Robotics Software Engineering RoSE, pp. 33–35. ACM, New York (2018)
6. Brambilla, M., Ferrante, E., Birattari, M., Dorigo, M.: Swarm robotics: a review from the swarm engineering perspective. Swarm Intell. **7**(1), 1–41 (2013)
7. Dimidov, C., Oriolo, G., Trianni, V.: Random walks in swarm robotics: an experiment with kilobots. In: Dorigo, M., et al. (eds.) ANTS 2016. LNCS, vol. 9882, pp. 185–196. Springer, Cham (2016). https://doi.org/10.1007/978-3-319-44427-7_16
8. Doncieux, S., Bredeche, N., Mouret, J.B., Eiben, A.E.G.: Evolutionary robotics: what, why, and where to. Front. Robot. AI **2**, 4 (2015)
9. Dorigo, M., Birattari, M., Brambilla, M.: Swarm robotics. Scholarpedia **9**(1), 1463 (2014)
10. Duarte, M., Oliveira, S., Christensen, A.: Evolution of hierarchical controllers for multirobot systems. In: Artificial Life Conference Proceedings 2014, pp. 657–664. MIT Press (2014)
11. Feynman, R.P., Leighton, R.B., Sands, M.L.: The Feynman Lectures on Physics Volume 1: Mainly Mechanics, Radiation, and Heat. Basic Books, New York (2011)
12. Francesca, G., Birattari, M.: Automatic design of robot swarms: achievements and challenges. Front. Robot. AI **3**(29), 1–9 (2016)
13. Francesca, G., et al.: AutoMoDe-Chocolate: automatic design of control software for robot swarms. Swarm Intell. **9**(2/3), 125–152 (2015). https://doi.org/10.1007/s11721-015-0107-9
14. Francesca, G., Brambilla, M., Brutschy, A., Trianni, V., Birattari, M.: AutoMoDe: a novel approach to the automatic design of control software for robot swarms. Swarm Intell. **8**(2), 89–112 (2014). https://doi.org/10.1007/s11721-014-0092-4
15. Garattoni, L., Francesca, G., Brutschy, A., Pinciroli, C., Birattari, M.: Software infrastructure for e-puck (and TAM). Technical report/IRIDIA/2015-004, IRIDIA, Université libre de Bruxelles, Belgium (2015)
16. Gutiérrez, Á., Campo, A., Dorigo, M., Donate, J., Monasterio-Huelin, F., Magdalena, L.: Open e-puck range & bearing miniaturized board for local communication in swarm robotics. In: IEEE International Conference on Robotics and Automation, ICRA, pp. 3111–3116. IEEE Press, Piscataway (2009)
17. Hasselmann, K., et al.: Reference models for Auto-MoDe. Technical report TR/IRIDIA/2018-002, IRIDIA, Université libre de Bruxelles, Belgium (2018)

18. Hasselmann, K., Robert, F., Birattari, M.: Automatic design of communication-based behaviors for robot swarms. In: Dorigo, M., Birattari, M., Blum, C., Christensen, A.L., Reina, A., Trianni, V. (eds.) ANTS 2018. LNCS, vol. 11172, pp. 16–29. Springer, Cham (2018). https://doi.org/10.1007/978-3-030-00533-7_2

19. Jakobi, N., Husbands, P., Harvey, I.: Noise and the reality gap: the use of simulation in evolutionary robotics. In: Morán, F., Moreno, A., Merelo, J.J., Chacón, P. (eds.) ECAL 1995. LNCS, vol. 929, pp. 704–720. Springer, Heidelberg (1995). https://doi.org/10.1007/3-540-59496-5_337

20. Kegeleirs, M., Garzón Ramos, D., Birattari, M.: Random walk exploration for swarm mapping. In: Althoefer, K., Konstantinova, J., Zhang, K. (eds.) TAROS 2019. LNCS (LNAI), vol. 11650, pp. 211–222. Springer, Cham (2019). https://doi.org/10.1007/978-3-030-25332-5_19

21. Kuckling, J., Ligot, A., Bozhinoski, D., Birattari, M.: Behavior trees as a control architecture in the automatic modular design of robot swarms. In: Dorigo, M., Birattari, M., Blum, C., Christensen, A.L., Reina, A., Trianni, V. (eds.) ANTS 2018. LNCS, vol. 11172, pp. 30–43. Springer, Cham (2018). https://doi.org/10.1007/978-3-030-00533-7_3

22. Ligot, A., Birattari, M.: On mimicking the effects of the reality gap with simulation-only experiments. In: Dorigo, M., Birattari, M., Blum, C., Christensen, A.L., Reina, A., Trianni, V. (eds.) ANTS 2018. LNCS, vol. 11172, pp. 109–122. Springer, Cham (2018). https://doi.org/10.1007/978-3-030-00533-7_9

23. López-Ibáñez, M., Dubois-Lacoste, J., Pérez Cáceres, L., Birattari, M., Stützle, T.: The irace package: iterated racing for automatic algorithm configuration. Oper. Res. Perspect. **3**, 43–58 (2016)

24. Miglino, O., Lund, H., Nolfi, S.: Evolving mobile robots in simulated and real environments. Artif. Life **2**(4), 417–434 (1995)

25. Mondada, F., Bonani, M., Raemy, X., Pugh, J., et al.: The e-puck, a robot designed for education in engineering. In: Gonçalves, P., Torres, P., Alves, C. (eds.) Proceedings of the 9th Conference on Autonomous Robot Systems and Competitions, pp. 59–65. Instituto Politécnico de Castelo Branco, Portugal (2009)

26. Floreano, D., Mondada, F.: Hardware solutions for evolutionary robotics. In: Husbands, P., Meyer, J.-A. (eds.) EvoRobots 1998. LNCS, vol. 1468, pp. 137–151. Springer, Heidelberg (1998). https://doi.org/10.1007/3-540-64957-3_69

27. Pasternak, Z., Bartumeus, F., Grasso, F.W.: Lévy-taxis: a novel search strategy for finding odor plumes in turbulent flow-dominated environments. J. Phys. A Math. Theor. **42**(43), 434010 (2009)

28. Pinciroli, C., Trianni, V., O'Grady, R., Pini, G., et al.: ARGoS: a modular, parallel, multi-engine simulator for multi-robot systems. Swarm Intell. **6**(4), 271–295 (2012)

29. Quinn, M., Smith, L., Mayley, G., Husbands, P.: Evolving controllers for a homogeneous system of physical robots: structured cooperation with minimal sensors. Philos. Trans. R. Soc. London A Math. Phys. Eng. Sci. **361**(1811), 2321–2343 (2003)

30. Ramachandran, R.K., Kakish, Z., Berman, S.: Information correlated Lévy walk exploration and distributed mapping using a swarm of robots. arXiv (2019)

31. Renshaw, E., Henderson, R.: The correlated random walk. J. Appl. Probab. **18**(02), 403–414 (1981)

32. Şahin, E.: Swarm robotics: from sources of inspiration to domains of application. In: Şahin, E., Spears, W.M. (eds.) SR 2004. LNCS, vol. 3342, pp. 10–20. Springer, Heidelberg (2005). https://doi.org/10.1007/978-3-540-30552-1_2

33. Silva, F., Duarte, M., Correia, L., Oliveira, S., Christensen, A.: Open issues in evolutionary robotics. Evol. Comput. **24**(2), 205–236 (2016)

34. Spaey, G., Kegeleirs, M., Garzón Ramos, D., Birattari, M.: Evaluation of alternative exploration schemes in the automatic modular design of robot swarms: Supplementary material (2020). http://iridia.ulb.ac.be/supp/IridiaSupp2020-002
35. Trianni, V.: Evolutionary Swarm Robotics. SCI, vol. 108. Springer, Berlin, Germany (2008). https://doi.org/10.1007/978-3-540-77612-3
36. Trianni, V., López-Ibáñez, M.: Advantages of task-specific multi-objective optimisation in evolutionary robotics. PloS One **10**(8), e0140056 (2015)
37. Watson, R., Ficici, S., Pollack, J.: Embodied evolution: distributing an evolutionary algorithm in a population of robots. Robot. Auton. Syst. **39**(1), 1–18 (2002)
38. Zaburdaev, V., Denisov, S., Klafter, J.: Lévy walks. Rev. Mod. Phys. **87**(2), 483–530 (2015)

Towards a Phylogenetic Measure
to Quantify HIV Incidence

Pieter Libin[1,2]($\boxtimes$), Nassim Versbraegen[5,6]($\boxtimes$), Ana B. Abecasis[2,3],
Perpetua Gomes[4], Tom Lenaerts[1,5,6], and Ann Nowé[1]

[1] Artificial Intelligence Lab, Department of Computer Science, Vrije Universiteit
Brussel, Brussels, Belgium
{pieter.libin,ann.nowe}@vub.ac.be
[2] Department of Microbiology and Immunology, Rega Institute for Medical Research,
KU Leuven - University of Leuven, Leuven, Belgium
[3] Global Health and Tropical Medicine, GHTM, Instituto de Higiene e Medicina
Tropical, IHMT, Universidade Nova de Lisboa, UNL, Lisbon, Portugal
ana.abecasis@ihmt.unl.pt
[4] Laboratorio Biologia Molecular, LMCBM, SPC, HEM, Centro Hospitalar Lisboa
Ocidental, Centro de Investigação Interdisciplinar Egas Moniz (CiiEM), Caparica,
Portugal
pcrsilva@chlo.min-saude.pt
[5] Machine Learning Group, Université Libre de Bruxelles, Boulevard du Triomphe
CP212, 1050 Bruxelles, Belgium
{nversbra,tlenaert}@ulb.ac.be
[6] Interuniversity Institute of Bioinformatics in Brussels, Université Libre de
Bruxelles-Vrije Universiteit Brussel, 1050 Brussels, Belgium

Abstract. One of the cornerstones in combating the HIV pandemic is
the ability to assess the current state and evolution of local HIV epidemics. This remains a complex problem, as many HIV infected individuals remain unaware of their infection status, leading to parts of
HIV epidemics being undiagnosed and under-reported. We first present
a method to learn epidemiological parameters from phylogenetic trees,
using approximate Bayesian computation (ABC). The epidemiological
parameters learned as a result of applying ABC are subsequently used
in epidemiological models that aim to simulate a specific epidemic.
Secondly, we continue by describing the development of a tree statistic, rooted in coalescent theory, which we use to relate epidemiological
parameters to a phylogenetic tree, by using the simulated epidemics.
We show that the presented tree statistic enables differentiation of epidemiological parameters, while only relying on phylogenetic trees, thus
enabling the construction of new methods to ascertain the epidemiological state of an HIV epidemic. By using genetic data to infer epidemic
sizes, we expect to enhance our understanding of the portions of the
infected population in which diagnosis rates are low.

Keywords: HIV incidence · Approximate bayesian computation ·
Phylogenetics · Coalescent theory

These authors contributed equally to this work.

© Springer Nature Switzerland AG 2020
B. Bogaerts et al. (Eds.): BNAIC 2019/BENELEARN 2019, CCIS 1196, pp. 34–50, 2020.
https://doi.org/10.1007/978-3-030-65154-1_3

1   Introduction

About 37 million people are currently infected with HIV and an estimated 35 million people have died due to the effects of AIDS (the eventual result of HIV infection) since the beginning of the epidemic at the start of the twentieth century [1]. Global efforts have ensued to enhance the collection, dissemination and accessibility of epidemiological data related to HIV. One of the most burdensome aspects in curtailing the spread of HIV emerges from infected individuals being unaware of their infection status. This stems from the fact that a host can be infected for many years before noticing any symptoms [2–4]. As a result, a significant fraction of the HIV infected population remains undiagnosed, hampering effectiveness of interventions and assessment of further developments of the epidemic. Consequently, methods that deliver a well-founded estimate of the number of HIV infected individuals are paramount [5]. Such an estimate provides insight regarding the number of undiagnosed infected individuals. State-of-the-art methods that aim to provide estimates of the size of HIV epidemics generally consist of applying compartment models to routine surveillance data to estimate the number of infected individuals (i.e. number of new diagnoses over time and $CD4^+$ cell counts) [6,7].

An abundance of clinical data is available in the context of HIV epidemics, as upon diagnosis a number of tests are performed and the results thereof collected. One of those tests determines the genotype of the virus infecting a patient [8]. To that purpose, the genetic sequence of the virus is determined. As a result, a vast number of HIV sequences have been collected over the last decades.

In this work we introduce a new method to quantify HIV incidence. The method relies on genetic data to gain insight into the specific sub-populations that contain a high rate of undiagnosed individuals, thus allowing for more effective health policies, through diagnosis strategies that are directed towards these particular sub-populations.

We validate our research on the HIV-1 epidemic in Portugal (Sect. 4.1). We therefore first present inference of the epidemiological parameters of said epidemic by applying approximate Bayesian computation [9]. We apply approximate Bayesian computation to fit a model that contains the epidemiological parameters in question (Sect. 4.2). We further show that calibrating simulations to specific epidemics is essential, as the epidemiological dynamics has an important impact on the shape of the phylogenetic tree (Sect. 5). We then construct a tree statistic that enables differentiation of epidemiological parameters based on phylogenetic trees (Sect. 4.4) and evaluate it on a set of epidemiological simulations (Sect. 5.3).

2   Background

2.1   Phylogenetic Trees

Phylogenetic trees are trees that capture evolutionary relationships between organisms. Figure 1 illustrates a phylogenetic tree's structure. A rooted tree consists of a root N_4, internal nodes $\{N_0, N_1, N_2, N_3\}$, leaves $\{L_0, L_1, L_2, L_3, L_4, L_5\}$ and branches interconnecting nodes with other nodes and leaves.

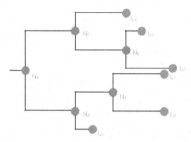

Fig. 1. An illustration of a phylogenetic tree

The horizontal branches of a phylogenetic tree indicate a measure of distance between the organisms represented by their respective leaves. This distance can be based on the amount of genetic change or can represent natural time, by using a molecular clock, i.e. an estimate of the time required for the observed genetic difference based on mutation rates [10]. A smaller path between two nodes, i.e. traversing the tree through the nodes from one leaf to another, thus indicates a stronger evolutionary relatedness.

Originally, such trees were constructed using morphological or phenotypic evidence [11]. Currently, genetic data has become the default data for phylogenetic tree construction [12].

Maximum Likelihood Methods. Maximum likelihood methods are widely used to construct phylogenetic trees. They rely on Maximum likelihood (ML); a statistical framework that employs an associated likelihood function [12]. The philosophy behind such methods consist of calculating the likelihood of hypothetical trees having produced the set of observed sequences, based on a model of sequence evolution [12].

However, calculating the likelihood with respect to all hypothetical trees would be intractable as the number of possible trees is $(2n - 5)!!$ with n the number of sequences [13, 14].

It has been shown that constructing a phylogenetic tree using maximum likelihood is NP-hard [14, 15]. Maximum likelihood implementations therefore rely on heuristic techniques to efficiently explore the search space. They thus explore a search space, where the tree currently being held as hypothesis has an associated likelihood value, and rearrangements of that tree that increase or decrease the likelihood. This approach alleviates the need to consider all possible hypothetical trees, by not exploring low likelihood regions unnecessarily. Maximum likelihood methods' main advantage resides in providing exact likelihoods, which allows for quantitative comparison of tree quality and can serve as an indication of the current position within the search space.

3   Human Immunodeficiency Virus

The Human Immunodeficiency Virus (HIV) is a retrovirus of about 120 nm. The HIV genome consists of nine genes, which code for 15 single proteins that are necessary for the virus to persist [16,17]. The most relevant ones in this context are envelope (*env*), group-specific antigen (*gag*) and polymerase (*pol*). *env* is especially relevant; it consists of *gp120* and *gp41*, which mediate the virus' interactions with the host's cells and are thus the main causative agent for HIV mainly targeting T-lymphocytes and macrophages expressing the $CD4^+$ antigen (i.e. specific parts of the human immune system) [18,19]. This mechanism results in a marked decline in the number of $CD4^+$ cells after HIV infection. After a period of latency (during which a specific equilibrium between HIV and $CD4^+$ cells emerges), the number of $CD4^+$ cells typically drops to low levels, while the number of HIV particles rises, rendering the infected host prone to opportunistic infections [20,21].

A single infected cell can produce thousands of new infectious HIV particles, either as an acute event, followed by cell death, or over a period of weeks, slowly releasing new HIV particles [16,22].

3.1   HIV Phylogenies

HIV evolves one million times faster than human DNA [23,24], mainly as a result of its error prone reverse transcriptase and the short lifespan of the viral generation [25–27]. As a result, the genetic variance among HIV viruses is large. Two types have been defined; HIV-1 and HIV-2 [28,29], both can result in acquired immune deficiency syndrome (AIDS), but HIV-2 progression to AIDS is generally slower [30].

HIV-1 has been divided in several groups; M, N, O and P. Group M is of particular interest, as it is the pandemic form, which has infected a vast amount of people [29,31]. Group O represents less than 1% of HIV-1 infections and is mostly found in west and south-east Africa [32]. Group N has an even lower prevalence, with only 13 documented cases up to 2010, all occurring in Cameroon. Group P has only been observed in two cases, and presumably accounts for 0.06% of HIV infection [33,34].

Group M has diverged into nine subtypes; A, B, C, D, F, G, H, J and K and over one hundred circulating recombinant forms (CRF) [29,35,36]. These are established through recombination of HIV strains, after an individual suffers infection from multiple subtypes (superinfection) and are considered CRFs if they are subsequently detected in three or more individuals who are not epidemiologically linked [37].

HIV-2 has at least eight distinct lineages (A-H), from which only groups A and B have significantly spread to humans [38].

Considering that infection by a specific subtype can result in different pathogenesis and resistance [39], it is essential to differentiate between subepidemics when conducting research on the topic. Tools have been developed to identify specific type and subtypes [40,41].

3.2 Epidemiological Models

Compartment models are one of the most popular concepts stemming from mathematical epidemiology [42], they aim to capture population dynamics by stratifying individuals into different compartments. As an example, the SIR model consists of 3 compartments; Susceptible (S), Infected (I) and Recovered (R) (or removed) and has transitions between those compartments (see Fig. 2). R, S and I represent the number of people in the respective compartment at a certain point in time. The rates β and γ are specific to infectiousness and pathology. The SIR model consists of three coupled non-linear ordinary differential equations, representing the change in each compartment over time;

$$\dot{S} = -\beta SI \tag{1}$$

$$\dot{I} = \beta SI - \gamma I \tag{2}$$

$$\dot{R} = \gamma I \tag{3}$$

With β the infection rate, γ the recovery rate and t time.

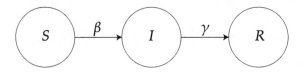

Fig. 2. SIR compartment model, compartments Susceptible (S), Infected (I) and Recovered (R) represented by circles, transitions between them by arrows, denoted by their associated rates (β and γ)

The model and extended versions thereof are especially relevant when trying to answer health policy questions. In our work, such models will be used to simulate specific HIV epidemics.

3.3 Detecting Missed Infections in Phylogenetic Trees

The starting point for our tree statistic was the method presented in [43], which uses the coalescent [44] to provide an indication of the extent to which samples are missing, or overly present in a specific phylogenetic tree.

Equation 4 specifies how to calculate node probabilities for each node j in a tree [43].

$$\pi^j(t, k, N) = \sum_{i=1}^{j} k_i^s Pr_i(C_{lineage}) \tag{4}$$

With i the interval index, t_i the interval length, k the number of lineages to coalesce in an interval, N the population size and $Pr_i(C_{lineage})$ the lineage coalescence probability.

$$Z^j \propto n\pi^j - np^j \tag{5}$$

To be able to gauge to what extent the population making up the tree under investigation is under- or oversampled, one can convert the node probabilities to z-scores, by using Eq. 5. This equation quantifies the relation between the number of descendant leaves expected at a node (i.e. $n\pi^j$) and the number of observed descendant leaves for that node (i.e. np^j). Descendant leaves are all the leaf nodes one obtains by following the branches of the node [43]. The z-score of a node thus reflects the extent to which more (or less) descendants are expected, given the observed descendants in the tree.

3.4 Portuguese HIV-1 Epidemic

The first reports of HIV-1 diagnoses in Portugal go back to 1983. By 2014, 53072 diagnoses had been reported [45]. As in most European HIV epidemics, subtype B is the most prevalent HIV-1 subtype, followed by subtype G, which is atypical in a European setting [46]. Within the "men who have sex with men" risk group, the number of yearly diagnoses show a mild but steady increase, while modes of transmission have transformed from intravenous drug use being the main cause of new infections to infections stemming from heterosexual sex in the period 2000–2014 [45]. This change is associated with lower diagnoses rates overall (consistent with diagnoses in Europe in general), possibly illustrating the beneficial results of a health policy implementation that was particularly effective for intravenous drug-users (e.g. through providing single-use needles). We apply our research on data stemming from this epidemic.

4   Methods and Materials

4.1 Portuguese Data

The data used in the experiments was made available through a HIV-1 resistance database from *Hospital Egas Moniz*. Henceforth, we will refer to the used data as 'Portuguese data' for the sake of brevity.

Data storage and querying was achieved through the RegaDB system [47]. Said system allows for complex querying, which was key to assemble all the relevant data in an efficient manner. We proceeded by querying genetic sequences of HIV-1 belonging to distinct patients. In doing so, we assembled three genetic sequence sets, differing in the HIV-1 subtypes they contain. The first set only holds subtype B sequences (n = 2216). The second set is made up of subtype G sequences (n = 1961). And a final set (n = 6079) that does not take into account the specific subtype (and thus also includes other subtypes besides B and G). HIV-1 sequences were classified using the Rega typing tool [48,49]. Each patient only contributed one genetic sequence to a set, if multiple sequences were associated with a single patient, the sequence established first was selected.

For each set, an alignment was created using MAFFT [50]. The resulting alignments were then used to infer phylogenetic trees using maximum likelihood trough RAxML [51]. RAxML was used with the GTR-γ model. A maximum likelihood tree was constructed and subsequently annotated through bootstraps. Bootstrapping was halted automatically based on extended majority rule consensus trees (i.e. autoMRE). In order to preserve the confidential nature of the employed patient data, tree inference was carried out on local computers exclusively and only anonymized patient data was used.

4.2   Simulation Calibration

In order to validate our tree statistic on a real world epidemic, being able to generate simulation data that was plausible with regard to the real world epidemic was essential. We thus proceeded by inferring relevant epidemiological parameters in order to calibrate subsequent simulations. To that end, we opted to use ABC [9] to learn said parameters, this approach was inspired by the work presented in [52]. ABC is closely related to Markov chain Monte Carlo (MCMC), but unlike MCMC, does not require the calculation of exact likelihoods, which can be intractable for complex models [52,53]. Learning the relevant epidemiological parameters in an ABC setting requires the presence of some distance measure, as an alternative to the exact likelihoods used in MCMC approaches. Taking into account that the available epidemic data we want to infer parameters from exists in the form of a phylogenetic tree, and the possibility of generating new phylogenetic trees through simulation, we employ a kernel method developed by Poon [54] as a distance measure between two trees. In concreto, we rely on a specified compartment model (see Fig. 3 for the used model) that enables the generation of trees.

By using the aforementioned kernel method, we assume that correspondence in trees reflect similarities between the model and the epidemic underlying the observed tree. In order to explore the possible parameters of the specified model efficiently, ABC is used. In essence, ABC varies parameter values in order to simulate more data by using the proposed parameter values in a specified model and aims to minimise the distance between the newly generated data and the observed data (in this case using the kernel method as a distance measure) [9,52]. Table 1 shows the parameters used in our ABC application. Broad parameter ranges were chosen around commonly used values in the field.

Table 1. Parameters used in ABC kernel method SI model, $X = e^{\mu+\sigma Z}$ represents the log-normal distribution, β_i being the infection rate, γ_i the mortality rate, μ the mortality rate from natural causes, N the population size.

Parameter	Range	σ	Initial	Prior
N	$[10^3–10^6]$	10^3	10^4	$X = e^{\mu+\sigma Z}$, $\mu = 0.5$ and $\sigma = 10000$
β_i	$[10^{-3}–10]$	0.05	0.5	$X = e^{\mu+\sigma Z}$, $\mu = 1.0$ and $\sigma = 0.01$
γ_i	$[0–5]$	0.01	0.1	$X = e^{\mu+\sigma Z}$, $\mu = 1.0$ and $\sigma = 0.01$
μ	$[0–1]$	0.002	0.02	$X = e^{\mu+\sigma Z}$, $\mu = 1.0$ and $\sigma = 0.01$

4.3   Simulations

Rcolgem [55,56] was used to simulate epidemics, based on parameter ranges obtained from application of the ABC-kernel method. Each simulation set consisted of 1000 simulations, outputting phylogenetic trees, and a log of the population dynamics over time. The used model (based on [57,58]) is given by the following equations, and is illustrated in Fig. 3.

$$\dot{S} = bN - \mu S - (\beta_0 I_0 + \beta_1 I_1 + \beta_2 I_2)\frac{S}{N}$$

$$\dot{I_0} = (\beta_0 I_0 + \beta_1 I_1 + \beta_2 I_2)\frac{S}{N} - (\mu + \gamma_0)I_0$$

$$\dot{I_1} = \gamma_0 I_0 - (\mu + \gamma_1)I_1$$

$$\dot{I_2} = \gamma_1 I_1 - (\mu + \gamma_2)I_2$$

The model is an extension of a SIR model, and consists of a susceptible state (S), three infection stages (I_0, I_1, I_2) and a deceased state (θ). The model includes births (determined by parameter b) and deaths (determined by parameter y_i) (i.e. conceptual addition and removal of simulated individuals) without a recovery state. Used parameter values, determined from the results of ABC application, are as follows; $\gamma_0 = 0.045$, $\gamma_1 = 0.14$, $\gamma_2 = 0.5$ $\mu = 0.001$, $\beta_0 = 0.12$, $\beta_1 = 0.03$ and $\beta_2 = 0.009$ while S_0 is varied between 145000 and 157000 and the sample size (i.e. the number of leaves in the tree) between 1000 and 12000. Parameters were sampled between the specified ranges using Latin hypercube sampling. The goal being to find parameters for the simulation engine that result in phylogenetic trees that are similar to the ones inferred from the Portugal data, in order to match the underlying epidemic.

4.4   Constructing the T_z-score

The starting point in constructing our tree statistic was the method described in [43]. The result of applying said method on a tree is an annotated tree, which includes node probabilities for each node in the tree. In order to construct our tree statistic from such an annotated tree, we devised a procedure to infer

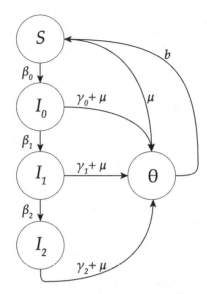

Fig. 3. Extended SIR compartment model, compartments Susceptible (S), three Infected stages (I_i) and Deceased (θ) represented by circles, transitions between them by arrows, denoted by their associated rates (infections β_i, death through infection γ_i, death trough natural causes μ and births b)

information about the population on the basis of z-scores. We proceeded by defining a statistic that demonstrates the overall extent to which a tree is over- or undersampled. We call this statistic T_z. To convert the obtained annotated tree to a single statistic value, we rely on the z-score in the root of the tree, as node probabilities are by definition propagated to the root node; $T_z = \frac{Z^r}{s}$, with Z^r the z-score of the root and s the number of samples (i.e. leaves) making up the tree. In order to apply the method, a tree and a N need to be specified. Through experimental analyses, we found that a large N is necessary to obtain informative results, we thus specified $N = 10^5$. Hence we obtain a single T_z score for each tree, allowing qualitative comparison between trees, while still being able to assess z-scores of individual nodes in the tree.

5   Results

5.1   Approximate Bayesian Computation

Figure 4 presents the results of ABC application in order to fit a specific tree. Here only the results for the parameter μ are displayed for clarity. The figure indicates that the ABC chain converged after about 1500 iterations and had thus learned plausible values for said parameters with regard to the specified model.

mu ABC result

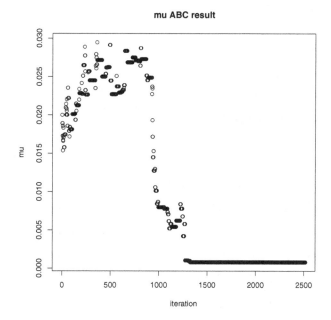

Fig. 4. ABC kernel method result, μ over different iterations

5.2 Phylogenetic Tree Assessment

We first present a visual comparison of tree topologies between the tree inferred from the Portuguese dataset, a tree obtained through an ABC application calibrated simulation and a PANGEA [59] simulation, which aims to model the HIV epidemic in sub-Saharan Africa. If topologies show major discrepancies, we assume this indicated the simulations are not well calibrated with regard to the actual epidemic, while a relatively corresponding topology would indicate simulations resembling the actual epidemic. The PANGEA tree serves as an example of topology difference when comparing different epidemics.

Figures 5a, 5c and 5e provide a visual representation of the relevant phylogenetic trees. Figure 5a stems from the Portuguese dataset, and as such offers a baseline of desired tree topology. Figure 5c presents the tree obtained through an ABC calibrated simulation, and Fig. 5e shows a tree from a PANGEA simulation. We demonstrate visually that the relevant tree topologies display a relatively high level of correspondence. We further investigated tree correspondence by using patristic distances [60]. The patristic distance between two leaves l_0 and l_1 is the number of changes needed to l_0 in order for it to become identical to l_1 [61]. Figures 5b, 5b and 5f present a histogram of the patristic distances present in the tree stemming from the Portuguese dataset the ABC calibrated simulated tree and the PANGEA tree respectively. These show that the ABC calibrated simulation tree is concordant with the tree stemming from the Portuguese epidemic.

(a) Cladogram of tree inferred from Portuguese dataset

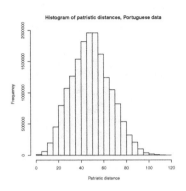

(b) Histogram of patristic distances in tree inferred from Portuguese dataset

(c) Cladogram of tree obtained through rcolgem simulation with simulation parameters based on ABC results

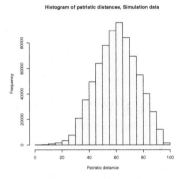

(d) Histogram of patristic distances in tree obtained through rcolgem simulation

(e) Cladogram of tree obtained through PANGEA simulation

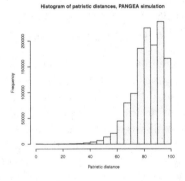

(f) Histogram of patristic distances in tree obtained through PANGEA simulation

Fig. 5. Cladogram of tree inferred from Portuguese dataset

5.3 T_z-score Distribution

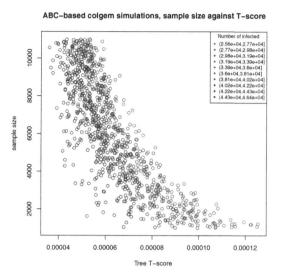

Fig. 6. Rcolgem simulation analysis, sample size (i.e. number of leaves) against T_z-score, with number of infected individuals in the simulations, shown as a gradient from 2.56×10^4 (red) to 4.64×10^4 (blue), $N = 10^5$ (Color figure online)

Figure 6 presents the result of application of our method on 1000 trees obtained from rcolgem simulations. In the figure, obtained T_z scores are plotted against tree sample sizes and the number of infected individuals in the simulations, shown as a gradient from 2.56×10^4 (red) to 4.64×10^4 (blue). As a reminder, the number of infected individuals is determined by β_i and S_0. We can clearly observe a distribution that allows distinction of the number of infected by T_z scores and sample sizes. Indicating that we obtained an informative distribution through application of our method. The figure shows that a lower T_z score correlates with a larger portion of the infected population not being included in the phylogenetic tree. Additionally, as sample size goes down the distribution becomes wider, indicating sufficiently large trees are necessary to allow for meaningful inferences.

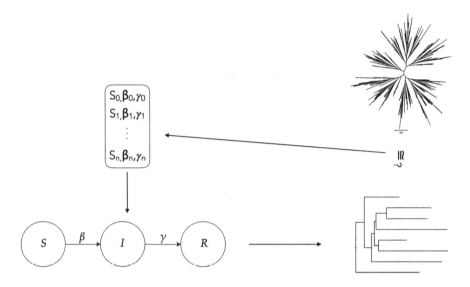

Fig. 7. ABC parameter calibration. A range of parameters is used in a compartment model to generate phylogenetic trees. These trees are then compared to a tree inferred from the actual epidemic data to fit the parameters.

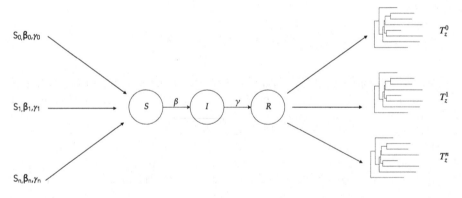

Fig. 8. T_z scores can be calculated on trees that were generated by compartment models with known epidemiological parameters, allowing the evaluation of said T_z scores.

6   Discussion and Conclusions

Figures 7 and 8 show a summary of the approach presented in this work. First, we can relate real-world genetic data to epidemiological parameters by applying ABC to an inferred phylogenetic tree (Fig. 7). Second, the development of the T_z score is enabled by the generation of trees with known epidemiological parameters, allowing us to relate T_z scores with said epidemiological parameters (Fig. 8).

The presented results show that the presented T_z score enables differentiation of epidemiological parameters based on phylogenetic trees. As such, an interesting further development would be to learn a function $f(S_0, \beta, \gamma, N) \rightarrow T_z^N$, i.e. learning the effect of epidemiological parameters on T_z values in general, and from there, $f(T_z^N) \rightarrow \{\beta, \gamma\}$, i.e. constructing a function that relates obtained T_z scores to possible epidemiological parameters, which we assume to be feasible, as S_0 should be ascertainable trough census data and epidemiological studies. We would further like to extend the T_z score to include a measure of uncertainty associated to inferences. A possible approach to accomplish this would be to apply our method on multiple subtrees, resulting from pruning the tree under investigation, and determining the extent to which results remain coherent with regard to the number of pruned leaves. Additionally, investigating the effect of simulating epidemics using current models specifically tailored to HIV-1 (e.g. an approach where the currently prevalent $CD4^+$ models would be adapted to generate phylogenetic trees) would be an interesting further development.

We have shown that application of the ABC method with compartment models allows us to capture trends of a real epidemic, but more complex models (such as individual-based models) might yield better correspondence with actual epidemic data. While the computational cost of such models should be considered, this would be an interesting avenue for future work.

Planned further research includes adaptation of the method presented in [43] to draw coalescent probabilities from a distribution that is specific to HIV evolutionary dynamics.

We have presented a tree statistic that can be employed to assort phylogenetic trees on the basis of their underlying epidemiological parameters. By doing so, we provide a first step towards a method to infer epidemiological parameters from phylogenetic trees using coalescent theory, which would additionally be able to indicate the specific subpopulations in which diagnosis rates are low, providing a crucial tool for health policy researchers.

Acknowledgments. Pieter Libin was supported by a PhD grant of the FWO (Fonds Wetenschappelijk Onderzoek Vlaanderen) and a grant of the VUB research council (VUB/OZR2714).

References

1. UNAIDS. Fact sheet - latest global and regional statistics on the status of the aids epidemic, July 2017
2. Hightow-Weidman, L.B., Golin, C.E., Green, K., Shaw, E.N.P., MacDonald, P.D.M., Leone, P.A.: Identifying people with acute HIV infection: demographic features, risk factors, and use of health care among individuals with AHI in North Carolina. AIDS Behav. **13**(6), 1075–1083 (2009)
3. Burton, G.F., Keele, B.F., Estes, J.D., Thacker, T.C., Gartner, S.: Follicular dendritic cell contributions to HIV pathogenesis. In: Seminars in Immunology, vol. 14, pp. 275–284. Elsevier (2002)

4. Theys, K., Libin, P., Pineda-Pena, A.-C., Nowe, A., Vandamme, A.-M., Abecasis, A.B.: The impact of HIV-1 within-host evolution on transmission dynamics. Curr. Opin. Virol. **28**, 92–101 (2018)

5. Begier, E.M., et al.: Undiagnosed HIV infection among New York City jail entrants, 2006: results of a blinded serosurvey. JAIDS J. Acquir. Immune Defic. Syndr. **54**(1), 93–101 (2010)

6. van Sighem, A., et al.: Estimating HIV incidence, time to diagnosis, and the undiagnosed HIV epidemic using routine surveillance data. Epidemiology **26**(5), 653 (2015)

7. Mammone, A., et al.: How many people are living with undiagnosed HIV infection? an estimate for Italy, based on surveillance data. AIDS **30**(7), 1131 (2016)

8. Vajpayee, M., Mohan, T.: Current practices in laboratory monitoring of HIV infection. Indian J. Med. Res. **134**(6), 801 (2011)

9. Sunnåker, M., Busetto, A.G., Numminen, E., Corander, J., Foll, M., Dessimoz, C.: Approximate bayesian computation. PLoS Comput. Biology **9**(1), e1002803 (2013)

10. Ridley, M.: Evolution. John Wiley & Sons Incorporated (2009)

11. Hennig, W.: Phylogenetic systematics. Ann. Rev. Entomol. **10**(1), 97–116 (1965)

12. Salemi, M., Lemey, P., Vandamme, A.-M.: The Phylogenetic Handbook: A Practical Approach to Phylogenetic Analysis and Hypothesis Testing. Cambridge University Press, Cambridge (2009)

13. Felsenstein, J.: Inferring Phylogenies, vol. 2. Sinauer associates Sunderland, MA (2004)

14. Chor, B., Tuller, T.: Finding a maximum likelihood tree is hard. J. ACM (JACM) **53**(5), 722–744 (2006)

15. Roch, S.: A short proof that phylogenetic tree reconstruction by maximum likelihood is hard. IEEE/ACM Trans. Comput. Biol. Bioinform. (TCBB) **3**(1), 92 (2006)

16. Freed, E.O.: HIV-1 replication. Somatic Cell Mol. Genet. **26**(1–6), 13–33 (2001)

17. Gallo, R., Wong-Staal, F., Montagnier, L., Haseltine, W.A., Yoshida, M.: HIV/HTLV gene nomenclature. Nature **333**(6173), 504–504 (1988)

18. Kindt, T.J., Goldsby, R.A., Osborne, B.A., Kuby, J.: Kuby immunology. Macmillan (2007)

19. Rambaut, A., Posada, D., Crandall, K.A., Holmes, E.C.: The causes and consequences of HIV evolution. Nature reviews. Genetics **5**(1), 52 (2004)

20. Cohen, M.S., Hellmann, N., Levy, J.A., DeCock, K., Lange, J.: The spread, treatment, and prevention of HIV-1: evolution of a global pandemic. J. Clin. Inv. **118**(4), 1244 (2008)

21. Fauci, A.S., Pantaleo, G., Stanley, S., Weissman, D.: Immunopathogenic mechanisms of HIV infection. Ann. Internal Medi. **124**(7), 654–663 (1996)

22. Ferguson, M.R., Rojo, D.R., von Lindern, J.J., O'Brien, W.A.: HIV-1 replication cycle. Clin. Lab. Med. **22**(3), 611–635 (2002)

23. Lemey, P., Rambaut, A., Pybus, O.G.: HIV evolutionary dynamics within and among hosts. AIDs Rev. **8**(3), 125–140 (2006)

24. Li, W.-H., Tanimura, M., Sharp, P.M.: Rates and dates of divergence between AIDS virus nucleotide sequences. Mol. Biol. Evol. **5**(4), 313–330 (1988)

25. Wei, X., et al.: Viral dynamics in human immunodeficiency virus type 1 infection. Nature **373**(6510), 117–122 (1995)

26. Ho, D.D., Neumann, A.U., Perelson, A.S., Chen, W., Leonard, J.M., Markowitz, M.: Rapid turnover of plasma virions and CD4 lymphocytes in HIV-1 infection. Nature **373**(6510), 123–126 (1995)

27. Holmes, E.C.: On the origin and evolution of the human immunodeficiency virus (HIV). Biol. Rev. **76**(2), 239–254 (2001)
28. Hahn, B.H., Shaw, G.M., De, K.M., Sharp, P.M., et al.: AIDS as a zoonosis: scientific and public health implications. Science **287**(5453), 607–614 (2000)
29. Sharp, P.M., Hahn, B.H.: Origins of HIV and the AIDS pandemic. Cold Spring Harbor Perspect. Med. **1**(1), 006841 (2011)
30. de Silva, T.I., Cotten, M., Rowland-Jones, S.L.: HIV-2: the forgotten AIDS virus. Trends Microbiol. **16**(12), 588–595 (2008)
31. Merson, M.H., O'Malley, J., Serwadda, D., Apisuk, C.: The history and challenge of HIV prevention. Lancet **372**(9637), 475–488 (2008)
32. Peeters, M., Gueye, A., Mboup, S., Bibollet-Ruche, F., Ekaza, E., Mulanga, C., Ouedrago, R., Gandji, R., Mpele, P., Dibanga, G., et al.: Geographical distribution of HIV-1 group O viruses in Africa. Aids **11**(4), 493–498 (1997)
33. Vallari, A., et al.: Four new HIV-1 group N isolates from cameroon: prevalence continues to be low. AIDS Res. Hum. Retroviruses **26**(1), 109–115 (2010)
34. Vallari, A., Holzmayer, V., Harris, B., Yamaguchi, J., Ngansop, C., Makamche, F., Mbanya, D., Kaptué, L., Ndembi, N., Gürtler, L., et al.: Confirmation of putative HIV-1 group P in Cameroon. J. Virol. **85**(3), 1403–1407 (2011)
35. Goudsmit, J.: Viral sex: The nature of AIDS. Oxford University Press on Demand (1997)
36. Buonaguro, L., Tornesello, M.L., Buonaguro, F.M.: Human immunodeficiency virus type 1 subtype distribution in the worldwide epidemic: pathogenetic and therapeutic implications. J. Virol. **81**(19), 10209–10219 (2007)
37. Taylor, B.S., Sobieszczyk, M.E., McCutchan, F.E., Hammer, S.M.: The challenge of HIV-1 subtype diversity. New Engl. J. Med. **358**(15), 1590–1602 (2008)
38. Santiago, M.L., et al.: Simian immunodeficiency virus infection in free-ranging sooty mangabeys (Cercocebus atys atys) from the Tai Forest, Cote d'Ivoire: implications for the origin of epidemic human immunodeficiency virus type 2. J. Virol. **79**(19), 12515–12527 (2005)
39. Baeten, J.M., et al.: HIV-1 subtype D infection is associated with faster disease progression than subtype A in spite of similar plasma HIV-1 loads. J. Infect. Dis. **195**(8), 1177–1180 (2007)
40. Pineda-Peña, A.-C., et al.: Automated subtyping of HIV-1 genetic sequences for clinical and surveillance purposes: performance evaluation of the new REGA version 3 and seven other tools. Infect. Genet. Evol. **19**, 337–348 (2013)
41. Libin, P., Versbraegen, N., Cuypers, L., Theys, K., Nowé, A.: An automated maximum likelihood method for classifying virus sequences (2016)
42. Brauer, F.: Compartmental models in epidemiology. Mathematical Epidemiology. LNM, vol. 1945, pp. 19–79. Springer, Heidelberg (2008). https://doi.org/10.1007/978-3-540-78911-6_2
43. Stacy, S., Black, A., Bedford, T.: Using the coalescent framework to detect missed infections in phylogenetic trees (2016)
44. Kingman, J.F.C.: The coalescent. Stochastic processes and their applications **13**(3), 235–248 (1982)
45. Diniz, A.: Portugal infeção VIH, SIDA e tuberculose em números, 2015. Portugal Infeção VIH, SIDA e Tuberculose em números, 2015, pp. 5–70 (2015)
46. Abecasis, A.B., et al.: HIV-1 subtype distribution and its demographic determinants in newly diagnosed patients in europe suggest highly compartmentalized epidemics. Retrovirology **10**(1), 7 (2013)
47. Libin, P., et al.: RegaDB: community-driven data management and analysis for infectious diseases. Bioinformatics **29**(11), 1477–1480 (2013)

48. Pineda-Peña, A.-C., et al.: Automated subtyping of HIV-1 genetic sequences for clinical andsurveillance purposes: performance evaluation of the new rega version 3 andseven other tools. Infection, Genetics Evol. **19**, 337–348 (2013)

49. Carlos, L., et al.: A standardized framework for accurate, high-throughput genotyping of recombinant and non-recombinant viral sequences. Nucleic Acids Res. **37**(suppl_2), W634–W642 (2009)

50. Katoh, K., Misawa, K., Kuma, K., Miyata, T.: MAFFT: a novel method for rapid multiple sequence alignment based on fast fourier transform. Nucleic Acids Res. **30**(14), 3059–3066 (2002)

51. Stamatakis, A.: RAxML version 8: a tool for phylogenetic analysis and post-analysis of large phylogenies. Bioinformatics **30**(9), 1312–1313 (2014)

52. Poon, A.F.Y.: Phylodynamic inference with kernel ABC and its application to HIV epidemiology. Mol. Biol. Evol. **32**(9), 2483–2495 (2015)

53. Libin, P., Hernalsteen, L., Theys, K., Gomes, P., Abecasis, A., Nowe, A.: Bayesian inference of set-point viral load transmission models. In: 30th Benelux Conference on Artificial Intelligence, pp. 107–121 (2018)

54. Poon, A.F.Y., Walker, L.W., Murray, H., McCloskey, R.M., Harrigan, P.R., Liang, R.H.: Mapping the shapes of phylogenetic trees from human and zoonotic RNA viruses. PLoS One **8**(11), e78122 (2013)

55. Rasmussen, D.A., Volz, E.M., Koelle, K.: Phylodynamic inference for structured epidemiological models. PLoS Comput. Biolo. **10**(4), e1003570 (2014)

56. Volz, E.M.: Complex population dynamics and the coalescent under neutrality. Genetics **190**(1), 187–201 (2012)

57. Volz, E.M.: Estimating HIV transmission rates with rcolgem (2014). Accessed Aug 2017

58. Volz, E.M.: Simulating genealogies with an epidemiological coalescent model using rcolgem (2015). Accessed Aug 2017

59. Ratmann, O., et al.: Phylogenetic tools for generalized HIV-1 epidemics: findings from the PANGEA-HIV methods comparison. Mol. Biol. Evol. **34**(1), 185–203 (2017)

60. Libin, P., et al.: PhyloGeoTool: interactively exploring large phylogenies in an epidemiological context. Bioinformatics **33**(24), 3993–3995 (2017)

61. Stuessy, T.F., König, C.: Patrocladistic classification. Taxon **57**(2), 594–601 (2008)

Cognitively Plausible Computational Models of Lexical Processing Can Explain Variance in Human Word Predictions and Reading Times

Wietse de Vries$^{(\boxtimes)}$

University of Groningen, Groningen, The Netherlands
wietse.de.vries@rug.nl

Abstract. Lexical processing times can yield valuable insights about structure in language and the cognitive processes that enable the use of language. Being able to estimate lexical processing times enables us to estimate readability and reading times of any text. It has been claimed that lexical processing times of words are influenced by word occurrence frequencies as well as the context it appears in (McDonald and Shillcock 2001; Baayen 2010). The context might be important because of predictive processes that enable quicker lexical processing (Christiansen and Chater 2016). In the present paper, the effects of morphosyntactic predictions on lexical processing times are investigated using two computational models. These computational models are trained to predict upcoming part-of-speech tags based on preceding part-of-speech tags and their predictions are compared with human predictions and human reading times from the PROVO corpus (Luke and Christianson 2018). A recurrent neural network is able to explain variance in human prediction errors whereas the Rescorla-Wagner model performs less well. The Rescorla-Wagner prediction associations do however explain more variance in human reading times. Moreover, the Rescorla-Wagner model associations explain more variance in gaze durations than human prediction errors. The human prediction errors and the Recorla-Wagner model associations combined explain most variance (Adj. $R^2 = 0.719$) in reading times, which indicates that the part-of-speech tag-based Rescorla-Wagner model associations contain complementary information to explicit human predictions about lexical processing times.

Keywords: Lexical processing · Cognitive modeling · Rescorla-Wagner · Recurrent neural networks · Reading · Language modeling

1 Introduction

Human brains are able to process information at great speeds, but more complex information requires more effort and processing time. Analysis of word-level lexical processing times can yield insights about what aspects of language influence

© Springer Nature Switzerland AG 2020
B. Bogaerts et al. (Eds.): BNAIC 2019/BENELEARN 2019, CCIS 1196, pp. 51–69, 2020.
https://doi.org/10.1007/978-3-030-65154-1_4

reading effort the most and this information can be useful at various levels of linguistic analysis. At the highest level, lexical processing time estimations can help to give more accurate full text reading times. At a lower level, lexical processing times can help to identify what parts of sentences may be hard to read, even though the sentence is considered to be grammatically correct. Finally, lexical processing times may be able to help explain why languages contain syntactic consistency and semantic redundancy. This paper describes the use cognitively plausible computational models to explain lexical processing times, but previous findings about lexical processing times will be described first.

1.1   Lexical Processing

Human lexical processing behaviour has previously been critical to the development of computational language models (Balota et al. 2007; McClelland and Rumelhart 1981). Reading times of words in isolation have for instance been studied to explain between-word differences and these reading times can yield a lot of information about the cognitive effort that is required for processing certain lexical features. Several factors have been claimed to explain variation in lexical processing times of individual words such as word frequency, concreteness and ambiguity, but lexical processing times can be better explained when context is taken into account (McDonald and Shillcock 2001; Baayen 2010).

McDonald and Shillcock (2001) introduce a concept called Contextual Distinctiveness that takes the frequencies of surrounding words into account to predict reading times. This context-dependent measure is able to explain more variance in lexical processing times than pure word frequencies (McDonald and Shillcock 2001). However, Baayen (2010) argues that the statistical approach of McDonald and Shillcock (2001) still only relies on the word frequency effect and this effect is only a epiphenomenon of learning word forms of lexical meaning. On the other hand, the Naive Discriminative Reader (NDR) model (Baayen et al. 2011) would be able to explain more variance in lexical processing times.

Baayen (2010) applies a frequency based language model and the NDR model on a data set with monomorphemic and monosyllabic words and concludes that the NDR model results explain most variance of human response latencies. The NDR model learns associations between nearby character unigrams and bigrams by applying the Rescorla-Wagner equations (Rescorla et al. 1972). The result is a two-layer connectionist network of which the weights are learned by iterating through raw text. The NDR model is ignorant about word order or semantics but analysis shows that the model still captures syntactic and morphological co-occurrence effects (Baayen 2010). The NDR model fully relies on character-level co-occurrence of words that are close to the target word to make word-level predictions which indicates that morphosyntactic patterns within words do influence lexical processing times. Baayen (2010) also states that lexical processing may also be influenced by higher level co-occurrence effects due to how words should be combined to create meaningful sentences. The influence of characters in preceding words on the lexical processing times of a future word suggests that a form of anticipation or prediction takes place before a word is processed.

Previous studies indicate that rather specific predictions are made before encountering linguistic input which enables humans to process spoken language quickly enough (Christiansen and Chater 2016). Through neural event-related potential studies it has been found that humans predict features like lexical categories (Hinojosa et al. 2005) and grammatical gender (Van Berkum et al. 2005).

The biases and patterns that influence lexical processing may not only be learned for individual words, but these factors may also apply to morphosyntactic features of these words (Baayen 2010; Luke and Christianson 2016). Therefore, the present paper attempts to explain lexical processing times based on morphosyntactic predictions. More specifically, computational models are trained to predict fine-grained word classes based on the word classes of context words. Example 1 illustrates this type of context-based word class prediction for the first sentence in the PROVO corpus based on the Penn Treebank part-of-speech tagset (Marcus et al. 1993).

Example 1. There are now rumblings that [...] $\equiv$ *EX VBP RB NNS IN*
Context: *EX VBP RB NNS*, Target: *IN*

Data for studying lexical processing times can be extracted from self-paced reading tasks or from eye-tracking data while reading. Especially eye-tracking data may be a good proxy for lexical processing time since eye-tracking does not make reading any more difficult than natural reading. As mentioned before, the present paper describes a prediction-based modeling approach for estimating lexical processing times. Word-level predictability scores can be estimated by letting humans predict upcoming words in a sentence using the cloze procedure (Taylor 1953). Predictability scores refer to the fraction participants that predict a target word correctly. This procedure lets participants of an experiment predict the next word in a sentence given the preceding words. Predictability of a word can be estimated by the fraction of correct predictions with a large group of participants. The PROVO corpus (Luke and Christianson 2018) is the largest English corpus that contain predictability scores and also includes eye-tracking data for the same sentences. Therefore, this corpus very valuable for evaluating the effects of context-based predictability on lexical processing.

Analyses on the PROVO corpus reveals that humans only predict 5% of content words correctly due to the sparsity of content words (Luke and Christianson 2016). As a result, the exact word predictability scores do not contain significant explanatory power for estimating reading times (Luke and Christianson 2016). The authors did however find that the lexical classes to which a word belongs are consistently predictable. It may for instance be hard to guess the next word in a sentence, but it may be clear that it is likely to be a noun. The lexical class-based predictability scores did significantly explain some variance in the reading times which makes the authors conclude that predictability of lexical features may influence reading times (Luke and Christianson 2016). The present paper will further investigate the effects of morphosyntactic predictability on reading times by using the PROVO data set. Computational models that should learn to predict morphosyntactic features are created and predictability scores from the

PROVO corpus can indicate whether the models are able to make similar predictions as humans even if the models perceive only a small set of morphosyntactic features as opposed to actual words. After this, it will be evaluated whether the model prediction outputs can explain human reading times. The results will then show whether the computational models mimic the prediction technique during lexical processing on a morphosyntactic level.

1.2   Computational Models

One type of computational model that is able to learn patterns in sequential data like language is the Recurrent Neural Network (RNN) (Elman 1990). A simple RNN with a single hidden layer can be used as a connectionist model for cognition. This type of network is able to make predictions based on the current input and the internal representation of the previous input. The connection to the previous hidden state enables the network to learn longer-distance dependencies between sequential inputs. It is not claimed that RNNs replicate human learning or comprehension (Elman 1990; Van Berkum et al. 2005), but RNNs have been applied successfully in linguistic tasks that involve pattern learning (Mirman et al. 2010; Misyak et al. 2010). Even though the RNN model is not designed as a cognitive model, its simple design can still be considered cognitively plausible. Therefore, the RNN architecture will be used for a computational model in the present paper.

A different kind of computational model that is designed specifically for modelling cognitive behaviour is the Rescorla-Wagner (R-W) model (Rescorla et al. 1972). This type of model learns associations between cues and outcomes based on error-driven learning. Associations between cues and outcomes are strengthened or weakened based on how well predictions match the outcomes. As a result, the model will not associate new cues with outcomes if a different cue already reliably predicts an outcome. A common example of this type of learning is Pavlov's dog. If a dog hears a bell before each meal, the bell will be associated with the meal. If after learning this, also a light will be flashed before each meal this will not be associated with the meal because the bell cue is already informative enough (Rescorla et al. 1972). This type of conditioning has been associated with behaviourist animal learning, but it has also been successfully used for modelling lexical processing (Baayen 2010). Therefore, this model is also used in the present paper.

RNN models and R-W models are expected to learn differently because of their different input representations, hidden state connections and general complexity. Due to the presence of one or more hidden layers in a RNN model and the lack of a hidden layer in a R-W model, a RNN model would be expected to make more accurate predictions. High accuracy is however not necessarily equivalent to human-like performance.

2   Method

To investigate the predictability of morphosyntactic features and the effects of morphosyntactic predictability, a RNN model and a R-W model are trained to predict part-of-speech (POS) tags of words based on the POS tags of preceding words. POS tag prediction is used instead of word prediction to eliminate the sparsity problem of words. The evaluation data is relatively small and does not give a good coverage of English words. As a benefit of generalising to POS tags the influence of interpretable morphosyntactic features can be evaluated.

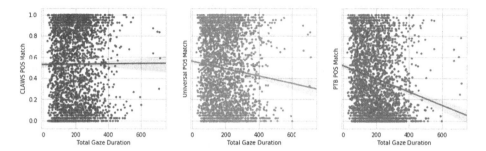

Fig. 1. Regression lines for correlation between human gaze duration per word for different types of POS match scores with a 95% confidence interval. POS match scores are the fractions of human word predictions that have the same POS tag as the target word.

The models have to be trained on a large amount of English texts so that the models will learn typical grammatical structures in English. Then, the models make predictions on unseen data. The models might rely on similar syntactic patterns as humans if model predictions explain variance in human prediction scores. Finally, it will be evaluated whether the models explain variance in human reading times. If this is the case, then lexical processing seem to rely on making morphosyntactic predictions. The differences between model results will give an estimation of the mechanism with which these predictions are made.

2.1   Data

Evaluation Data. For comparison between model predictions and human predictions and reading times, the previously discussed PROVO corpus (Luke and Christianson 2018) is used. However, this corpus contains just 55 texts with 132 sentences in total. This small amount of data does not cover the large amount of syntactic variation that is present in natural language. Additionally, the models should not be trained on this evaluation data if a fair comparison between humans and models should be made. The human prediction and reading time measures orginate from humans who have not read these particular texts before but, since they are humans, they have read large amounts of other texts before participating in the experiment. Because of this, the PROVO corpus is only used for evaluation while model training is executed with a large set of different texts.

The texts in the PROVO corpus are a combination of excerpts of news articles, popular science articles and fiction (Luke and Christianson 2018). On average, the text excerpts contain 2.5 (range 1–5) sentences of 13.3 (range 3–52) words per sentence. Word predictions are collected for all words except the first word of each document using a cloze-procedure experiment with 470 university students. The documents are presented to the participants one word at a time and the participants have to type in their prediction of the next word at each step. As a result, each word in the PROVO corpus contains at least 40 predictions. Reading time data is collected using an eye-tracking experiment with 84 participants that did not participate in the prediction experiment (Luke and Christianson 2016). During reading, humans tend to move back and forth between words. As a result, the eye-tracking data contains multiple measures like the first fixation duration and the amount of fixations on the target word (Luke and Christianson 2018). The first fixation duration may be informative, but the total time a human spent looking at a word may be a better indicator for lexical processing time. Therefore, the total fixation duration, the gaze duration, will be used in this paper as a measure for lexical processing times. Gaze durations that are shorter than 80ms or longer than 800ms have been removed in the creation of the corpus (Luke and Christianson 2018). As a result, gaze duration times per word are close to normally distributed. Therefore, the mean gaze durations per word are used in this paper as a proxy for human lexical processing times.

Training Data. The data that is used to train the language models is the WSJ part of the Penn Treebank (PTB) (Marcus et al. 1993). This annotated corpus of news articles does not cover all genres of text present in the PROVO corpus, but the texts in both corpora are written by professional authors. Therefore, the sentences in both corpora are likely to be considered to be grammatically correct according to common English writing standards. This similarity is important since the models have to fully rely on syntactic structure without additional semantic cues that are present in informal spoken and written language. The PTB data set consists of 49,208 sentences from 2312 news articles. For model efficiency purposes, only sentences with 3 to 60 tokens are used. This removes only 1.0% of the sentences in the PTB.

Part-of-Speech Tags. Both the PTB and the PROVO corpus contain POS tag information for each word but different tag sets are used in each corpus. The PTB corpus contains human annotated POS tags using its own tag set that is commonly used. The POS tags in the PROVO corpus originate from the automatic CLAWS tagger (Garside 1997). Due to obvious tagging mistakes and incompatibility between the PTB tag set and the CLAWS tag set, the PROVO data is re-tagged to the PTB tag set. This re-tagging is done using the default NLTK POS tagger due to its decent performance (Loper and Bird 2002). The PTB tag set contains tags for lexical classes like verbs and nouns, but classes are also separated by morphosyntactic features like tenses and plurality. As a result, the tag set contains 36 word-level POS tags and 9 additional tags for different kinds of punctuation. In order to investigate the differences between lexical categories without distinction between finer-grained categories, the PTB POS tags are converted to the Universal tag set (Nivre et al. 2018) using the official conversion table[1]. This simpler tag set contains 11 unique tags that correspond to lexical classes. The computational models are however trained on the more fine-grained PTB tags. Figure 1 illustrates a preliminary view of how POS match scores of different tag sets correlate with gaze durations. POS match scores are the fractions of human predictions that have the same POS tag as the target word. The figure illustrates that the CLAWS match scores that are based on a tag set that is nearly the same as the Universal tag set does not correlate with the gaze durations. The Universal POS match scores do correlate with gaze durations and the effect is even stronger for the finer-grained PTB POS tags. This correlation suggests that morphosyntactic predictability may influence gaze durations, but the correlation may for instance also be caused by different word length distributions between POS tags. Figure 2 illustrates that not all word classes are equally predictable. Note that proper nouns are never correctly predicted by humans. A possible reason may be that participants in this experimental setting avoid guessing names because the chance of guessing a proper name is too small in general.

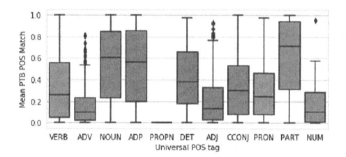

Fig. 2. Distributions of PTB POS Match scores per Universal POS tag. Boxes indicate quartiles, whiskers the 1.5 interquantile range and dots indicate outliers.

2.2   Recurrent Neural Network Model

Training: The first model that is trained to predict POS sequences is a fully connected simple recurrent neural network (RNN). The input layer and output layer each contain 45 input units that correspond to each possible PTB POS tag and the recurrent hidden layer contains 300 units. Figure 3 illustrates the RNN network structure. At each time step, the input unit is activated that corresponds to the POS tag of the previous word and the target output is the

---

[1] https://universaldependencies.org/tagset-conversion/en-penn-uposf.html.

POS tag of the next word. To get output unit activations that correspond to POS tag probabilities, the softmax function is applied to the output layer.

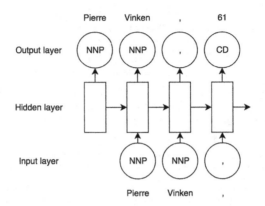

Fig. 3. RNN network structure. Each circle represents 45 fully connected units and each box represents 300 fully connected units. For each target, the preceding POS tag is used as input. Identity of earlier POS tags is derivable through access to the preceding hidden state.

Hyper-parameters for the network are chosen based on manual tests to improve accuracy without overfitting. Because of the risk of overfitting in loss minimizing neural networks, 10% of the data is set aside as validation data. The network is trained for 50 epochs using the Adam optimizer (Kingma and Ba 2014) with learning rate 0.001 and the categorical cross-entropy loss function. The order of the sentences is shuffled between each epoch so sentences that originated from the same source text are not processed in order. A dropout rate of 20% is set between the recurrent layer and the output layer to prevent overfitting (Srivastava et al. 2014). As a result, the training and validation errors are very similar throughout the training process.

The trained model has achieved a categorical prediction accuracy on the PTB training data of 38%. This accuracy score includes punctuation predictions and does not yield any information about certainty and biases. Generally, the trained model also predicts less common POS tags on the training data, but rare POS tags do not always occur in predictions.

Testing: For comparison with human predictions, the PROVO sentences are passed through the frozen network without allowing any weight updates to take place. The predictions where the target tag is punctuation are removed to match the human experimental results. Punctuation outputs are also removed since predicting punctuation was not an option for humans and the outputs are rescaled to sum up to one again. For comparison with human results, multiple measures are extracted from the model prediction outputs: the target value activation, the maximum activation value (response activation) and the mean squared error (MSE). These three values are closely related but they represent different kinds predictability information. The *target activation* represents how probable the actual POS tag is, whereas the *response activation* represents the extent to which any specific POS tag is expected. Finally, the *MSE* gives an overall error measure of the possible outputs. Due to the nature of how the MSE is calculated, this measure emphasizes the distribution of non-target outputs. In other words, the MSE is high if the absolute prediction error is equally distributed over the different non-target outcomes.

2.3   Rescorla-Wagner Model

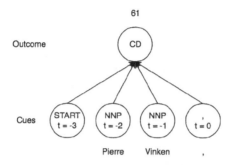

Fig. 4. Rescorla-Wagner network structure with temporally encoded cues.

Training: The Rescorla-Wagner (R-W) model is not designed to process sequential input like the sentences that are to be processed. Architecturally, the R-W model is equivalent to a perceptron model excerpt that a different learning strategy is used. To deal with the sequential nature of language, positional encoding is used for cues. The POS tag of each word in the PTB corpus is used once as the outcome for the R-W model with all preceding word POS tags as its cues. This network structure is illustrated in Fig. 4. The cues are all suffixed with its distance from the outcome. This means that a single word serves as a unique cue for each upcoming word. This positional encoding enables the model to be associated with multiple outcomes at different positions. A determiner may for instance be associated with an adjective at the next position and with a noun at the position after that without any conflict. A shortcoming of this approach is that all cues are presented with the same salience, even though some cues may represent words that were read up to 60 words before. The RNN model is able to selectively remember and forget information about preceding POS tags, but the R-W model cannot do this. Figure 5 does however illustrate that this may not be a problem. Associations with POS tags at longer distances are not learned because they are not consistently present in different sentences. Mean association strengths therefore decrease quickly as distance increases.

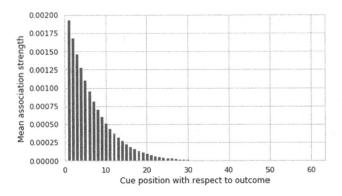

Fig. 5. Mean Rescorla-Wagner association strengths per distance between cue and outcome.

In addition to the position encoded POS tags corresponding to the preceding words in the sentence, a special *start* cue is added with its relative position. This allows the network to take the position of the outcome in the sentence into account. This is added to let the model know whether it is predicting the beginning or end of a sentence.

The α parameter of the R-W model is set to 0.00001 and all generated cues and outcomes of the PTB corpus are passed through the model in random order to learn associations. Only a single iteration is used since multiple iterations through the full data set do not improve prediction accuracy significantly. Additional iterations would increase the size of differences between associations, but this does not affect outcomes. For the same reason, the α parameter is set to a low value. Larger values up to 0.01 do not significantly change prediction accuracy but higher values lead to a sudden increase in training accuracy and a decrease in validation accuracy. The choice of a small α parameter prevents this kind of overfitting and retains most nuances between associations of different possible outcomes.

Testing: After the single learning iteration through the PTB data set, predictions for the PROVO corpus sentences are extracted from the model without updating the associations. For comparison with human data, the association with the target POS tag (target activation) as well as the highest association value (response activation) are extracted for each word in the corpus.

3   Categorical Results

Before the evaluation of how well the RNN model and the R-W model match human prediction behaviour, the categorical predictions of the words are analyzed. To do this, the most likely predictions per word are compared to get accuracy values for the models and for humans. These predictions are collected

Table 1. Accuracy scores for humans and models on PTB and Universal tag sets.

	PTB Acc.	Universal Acc.
Human	0.482	0.561
RNN	0.372	0.459
R-W	0.321	0.383

for both the PTB tag set and the Universal tag set to analyze more fine-grained differences. Table 1 shows that the two computational models do not outperform human predictions based on overall accuracy. Accuracy scores may seem low, but all scores are much higher than would be expected from random guessing. Therefore, both models as well as the humans must take POS-tag information into account. The RNN model seems to make more accurate predictions overall than the R-W model. However, the absolute accuracy difference between the RNN model and the R-W model is 5.1% for the PTB tag set whereas it is 7.6% for the Universal tag set. This suggests that the RNN model predicts a similar POS tag more often than the R-W model when the exact prediction is incorrect. In other words, the RNN model may for instance generalize over verbs and predict a singular verb where a plural verb is the target. This kind of generalization is not possible in the R-W model due to the simplicity of the model design.

The reason that humans outperform the models is most likely because native English speaking adults have read more text than is provided to the models as training data. Additionally, the humans received actual words as context instead of POS tags and there may be more informative semantic clues than only the information captured by the POS tag. Nevertheless, accuracy scores of both models are close enough to human performance for it to be possible that the models make similar generalizations as humans do. Such a conclusion can however not be made by looking at accuracy scores alone, since scores cannot be compared directly. The models received less information than the human participants so if model scores have predictive power for human scores, that suggests that the subset of information that the models get is actually an informative subset.

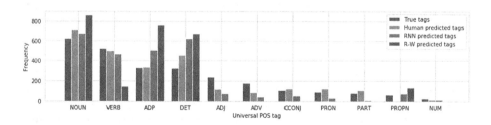

Fig. 6. Actual or predicted POS tag frequencies for humans and models. Actual PTB POS tags are aggregated per Universal POS tag for clearer interpretation, but the distribution of fine-grained PTB POS tags is equivalent.

Figure 6 shows that humans and the models both have prediction biases towards certain POS tags relative to actual frequencies. This figure shows for instance that humans generally expect nouns and determiners more often than their actual occurrence frequencies whereas verbs are predicted less often. The model responses, and especially the R-W model responses, have larger deviations from the target frequencies but the direction of the deviation always matches the over- or underestimation direction of humans. The only exception to this rule are proper nouns since humans never predicted a proper noun in the experiment. This does not have to mean that proper nouns as a word category are always unexpected. It is more likely that they do not to randomly guess names as opposed to actual English words since there is an unbound number of possible names. In summary, the frequencies suggest that the models have biases in a similar direction as humans but the effects are much stronger in the models. Mainly the shared underestimation of verbs suggests that the models do not just predict the most likely POS tag based on frequency since verbs occur frequently. Therefore, the frequencies suggest that the models and humans might share similar generalizations. This can however not be concluded based on raw frequencies, so therefore the explanatory power of model predictions for explaining human predictions will now be evaluated.

4   Computational Models

4.1   Modelling Human Predictions

If the computational models make predictions similarly to humans, then the model results should have explanatory power for predicting the fractions of predictions by humans that correspond to the target POS tag (human POS match scores). Therefore, statistical models are created with the human POS match scores as the dependent variable. The values of this dependent variable range between zero and one with a skewed distribution with many values close to zero. Nevertheless, an linear ordinary least squares model is sufficient to model the data since the final models fit the data nicely with nearly normally distributed residuals and no violation of assumptions. An advantage of these linear models over more advanced modeling techniques is that the directions of the estimator effects are easily evaluated. Models are created by stepwise addition of estimators and interactions if they improve the AIC score. The models that fit the human POS match scores the best are shown in Table 2.

Baseline Model. Before computational model outputs are used as estimator for human match scores, a baseline model is created which has to be improved by adding the model estimators. The baseline model contains the POS tag and the word length as estimators where the word length is the amount of characters in the original word. The effects in the resulting model for instance reveal that humans are inherently better at predicting POS tags like singular nouns ($p <$ 0.001) than adverbs ($p < 0.001$). The baseline model also indicates that humans

Table 2. Ordinary least squares (OLS) model results for human POS prediction accuracy per prediction model. *Baseline* is used as a stand-in for *POS * WordLength*.

Model	OLS estimators	PTB tag set		Universal tag set	
	(Target: human prediction accuracy per word)	AIC	Adj. R^2	AIC	Adj. R^2
WordLength	*WordLength*	1494.025	0.017	1802.648	0.002
POS	*POS*	890.363	0.231	887.159	0.308
Baseline	*POS + WordLength*	873.576	0.236	882.760	0.310
RNN	*Baseline + POS * TargetAct + MSE : TargetAct*	**622.407**	**0.315**	**594.474**	**0.390**
R-W	*Baseline + POS * TargetAct + ResponseAct*	713.222	0.290	680.430	0.369
RNN + R-W	*RNN + R − W*	643.857	0.317	591.216	0.397

are somewhat better at predicting shorter words than longer words because the effect of word length is negative ($p < 0.001$). The usage of POS tags as estimators is important since match scores may be different for different POS tags due to biases and frequency differences. The computational models should however be able to add additional information. The word length is by itself not a good estimator for human prediction accuracy but in combination with the POS estimator it does significantly contribute to the model. Shorter target words appear to be easier to predict by humans whereas the POS tag based computational models cannot know whether the original target word was short. The residuals of the baseline model are not normally distributed, which suggests that there may be structure in the data that is unaccounted for.

RNN Model. An ordinary least-squares model is created for the RNN model using the same procedure that was used for the baseline model. In addition to the POS and word length estimators, the RNN target activation variable significantly improves the model fit for both the PTB POS match as well as the Universal POS match target variables. The effect is positive ($p = 0.001$), which indicates that the RNN target activation significantly predicts human responses. In addition to the main effects, the interaction between the POS tag and the target activation improves the model fit as well as the interaction between the target activation and the MSE value. There is a strong positive effect for the interaction between target activation and MSE ($p < 0.001$) whereas the main effect of MSE does not significantly improve the model. The interaction indicates that the POS match scores will be high when the target activation is high while the MSE is also high. This is a curious interaction because a high target activation value should correspond with a low error, but the MSE will also be high if the activations are evenly distributed over the possible outcomes. Therefore, this interaction indicates that the POS match is high if the target POS tag is relatively probable while all other POS tags are all equally improbable.

R-W Model. Similarly to the RNN model, an ordinary least-squares model is created for the R-W using stepwise addition of estimators on top of the baseline model. The target activation value adds most explanatory power to the model with a strong positive effect ($p < 0.001$). This indicates that the R-W association is a good estimator for human predictions. In addition to the main effect for target activations, a main effect for the response activations also significantly contributes to the model with a negative effect ($p = 0.003$). As a result, the POS match is predicted to be lower if the response activation is high. If the response is equal to the target, the POS match will still be high because the coefficient of the target activation is much higher. In addition to these main effects, an interaction between the POS tag and the target activation is added to the model, because the target activation effect is not equally strong for all POS tags.

Model Comparison. The AIC scores and the adjusted R^2 scores are shown in Table 2 for each OLS model described in the previous sections. These results clearly show that both the RNN model and the R-W model reveal predictive information for human predictions that was not present in the baseline model. The table also shows that the RNN model predicts human responses better than the R-W model. This indicates that the more complex structure of the RNN model and the loss minimizing learning procedure do not only lead to better prediction accuracy but predictions also match human predictions more closely. It is not likely that the RNN model and the R-W model contain complementary predictive information since a union of the RNN and the R-W estimators does not improve the model fit. These findings show that predictions of the computational models and especially the RNN model at least partially depend on the same patterns that humans use. Therefore, these results suggest that syntactic patterns influence human predictions. This is not an entirely surprising conclusion since humans have an intuition whether a sentence is grammatically correct, but this at least reveals that the computational models are able to learn the same grammatical patterns.

4.2   Modelling Human Gaze Durations

The human prediction model results show that the computational models are able to reveal information about how humans explicitly predict upcoming words in a sentence. In this section the gaze durations are modelled using the prediction variables of the computational models. The influence of POS predictability on gaze durations is investigated in this section, so therefore the human POS match scores are also used as a estimator in the statistical models. If the influence of syntactic predictability on gaze durations is very strong and if the models approximate human cognition very well, it would be theoretically possible that computational model results reveal more predictive information than the POS match scores. The results of the best ordinary least squares models are listed in Table 3.

Table 3. Ordinary least squares (OLS) model fits and explained variance for gaze durations per word. The *Baseline, Human, RNN* and *R-W* estimators are stand-ins for the estimators in their respective models with the exclusion of duplicates.

Model	OLS estimators (Target: human gaze durations per word)	AIC	Adj. R^2
POS	*POS*	2560.335	0.515
WordLength	*WordLength*	2009.570	0.604
Baseline	*POS + WordLength*	1366.281	0.695
Human	*Baseline + PTB_POSMatch*	**1278.411**	**0.706**
RNN	*Baseline + POS * TargetAct*	1333.473	0.702
R-W	*Baseline + POS * TargetAct + POS * ResponseAct*	**1282.181**	**0.711**
RNN + R-W	*RNN + R − W*	1309.381	0.711
Human + RNN	*Human + RNN*	1267.892	0.710
Human + R-W	*Human + R − W*	**1215.665**	**0.719**

Baseline Model. A baseline ordinary least squares model is created to evaluate whether the computational model predictions actually add new information. The dependent variable, the gaze duration distribution is skewed with relatively many long values due to the temporal nature of the variable. To make the distribution more close to normal distributed, the gaze durations are log-transformed. Gaze durations are modelled using the same baseline features that were used to predict the POS match scores: the POS tag and the word length. A model is selected using the same stepwise addition procedure that was used before. The best model contains main effects for the POS tag as well as the word length. In addition to these main effects, the interaction between POS tag and word length would significantly improve the model fit, but this is likely caused by inherently different distributions of word lengths between POS tags. The interaction is excluded from the model because the model would overfit to specific words for some POS tags. The baseline model is already a decent estimator for gaze duration because it already explains 70.6% of the variation in the gaze durations according to the adjusted R^2 value. Neither the POS nor the word length estimators predict this much variance as sole estimators which indicates that both estimators are important.

Human Model. To evaluate whether POS predictability influences gaze durations, a statistical model is created to predict gaze durations using human POS match scores. The predictive power of the calculated PTB POS and Universal POS match scores are tried as estimators as well as the CLAWS POS match scores that were calculated by Luke and Christianson (2018). Of these three variants of POS predictability, the PTB POS match scores explain most variance. The addition of other POS match variants nor the addition of interactions improve the model fit. This indicates that the PTB tag set approximates cognitive symbolic representations of syntactic labels most closely.

RNN Model. Now a similar statistical model is created with RNN model outputs to evaluate whether the RNN model reveals information about gaze durations. In addition to the baseline estimators, the target activation is added as a main effect as well as an interaction between the POS tag and the target activation. The main effect of the target activation is negative ($p = 0.04$) which indicates that gaze durations are shorter if the target is predictable by the RNN model. The response activation nor the MSE estimators do not significantly improve the model fit. This suggests that the gaze duration is not significantly influenced by the probability distribution of non-target responses.

R-W Model. Of the candidate R-W estimators, the response activation improves the baseline model fit with the largest effect although the target activation is a close second. Stepwise addition reveals that the use of both the target activation and the response activation main effects improve model fit as well as the interactions of both estimators with the POS tags. In this model, the gaze duration significantly decreases with both higher target activations ($p = 0.04$) as well as higher response activations ($p < 0.001$). This indicates that gaze durations are shorter when POS tags are predictable in the R-W model.

Model Comparison. Table 3 shows how well the best versions of each predictive model are able to explain variance in gaze durations. The model with human PTB POS match scores fits the gaze durations significantly better than the baseline model, but the two computational models do too. According to the AIC scores, the models for RNN and R-W fit the data less well than the human POS match based model but more noteworthily, the R-W model seems to fit better than the RNN model. This reversal of positions with respect to the POS match prediction model may have revealed some important information about nature of these algorithms with respect to cognition because the models seem to match different tasks. Maybe more importantly than the AIC score comparison, the results show that the R-W model is able to explain just as much information as the human POS match based model. The adjusted R^2 value is even slightly higher, but the difference is really small.

To investigate whether different models contain complementary information about gaze durations, the estimators of the computational models are combined. Table 3 shows that the combination of the RNN model and the R-W model does not improve model with respect to single models. The combination of human POS match scores and the RNN estimators neither improves the model which indicates that the RNN model does not reveal any additional information. It does however appear that the union of the human prediction model and the R-W model significantly improves model fit ($\Delta AIC = 62.746$). Additionally, the union model also explains more variance according to the adjusted R^2 measure. This result indicates the associations learned by the R-W model are able to explain variation in gaze durations that could not be explained by the human prediction data. Therefore, it can not only be concluded that gaze durations are influenced by syntactic predictability and that the models are able to make similar predictions, but the results show that the R-W model associations explain additional information that could not be explained by explicit human predictions since the R-W model explains additional variance to the human prediction scores.

5   Conclusion

This paper started with the premise that lexical processing times are influenced by the predictability of syntactic patterns. Both the RNN and the R-W should be able to make accurate predictions to some extent and due to the added complexity of the RNN model, the RNN model would likely have a higher accuracy. The results have indeed shown that this is the case. The accuracies do however not indicate whether the models make the same mistakes as humans nor does it indicate whether the models are certain about the same words that have high predictability agreement by humans.

Table 2 shows that both models are able to explain variance in human predictions on top of the variation explained by POS tags and word length. However, the RNN model fits the human prediction data much better and explains more variance than the R-W model. The results show that the RNN model explains most variance for human prediction accuracies with respect to the R-W model. Therefore the cognitive process of humans may have similar features as the RNN model for the task of next word prediction. Additionally this suggests that humans use morphosyntactic features of preceding words to make predictions.

The results for the gaze duration models are however quite different as can be seen in Table 3. These results show that the human predictions have some explanatory power for gaze durations and therefore for lexical processing time. Noteworthily, the R-W model results fit the human gaze durations better than the RNN results and the R-W model results explain even a little more variance than the model based on human prediction accuracies. This suggests that the R-W model may be a good approximation of the cognitive process that takes place when humans read text. The model was trained on relatively little data with only the ordered POS tags as input and output. Therefore, it seems that human lexical processing times depend on morphosyntactic predictions. When combining the human POS matches with the R-W estimators, even more variation of gaze durations is explained. Therefore, some associations found by the R-W model are actually better estimators of lexical processing time than human POS predictions.

The main question of this paper was whether word-level syntactic predictability plays an important role in lexical processing and two computational models were compared. It appears that the RNN is the best model to explain human predictions. The reason for this is likely that the RNN architecture allows more complexity in finding the solution and the human brain is capable of making these complex predictions when thinking consciously. Lexical processing while reading text is also influenced by syntactic predictability, but it appears that this does not work with the exact same cognitive system that is used for conscious predictions. The gaze durations are better explained by the simple associations in the R-W model, which means that humans require a longer lexical processing time if certain morphosyntactic properties are not anticipated. Explicit predictions and gaze durations are both influenced by a prediction mechanism, but the differences between models suggest that the associated cognitive processes are not the same.

The results that are described in this paper may have strong implications in cognitive linguistic research because more information about the cognitive process of reading text is revealed. However, these results may also impact readability research and applications. The R-W model that is built for this paper can be applied on unseen text to give a numeric indication about readability if one assumes that reading pace is a good proxy for readability. Existing readability formulas and measures are based on heuristics and content whereas the R-W model is data driven and able to model lexical processing times. To demonstrate this type of application, a demo is created that calculates gaze durations for any text based on the R-W model outputs and the coefficients of the statistical model. The demo visualizes the adjustments of reading times that are explained by the R-W model estimators. The demo is publicly available at https://lexical-processing.wietsedv.nl.

References

Baayen, R.H.: Demythologizing the word frequency effect: a discriminative learning perspective. Mental Lexicon **5**(3), 436–461 (2010)

Baayen, R.H., Milin, P., Durdjevic, D.F., Hendrix, P., Marelli, M.: An amorphous model for morphological processing in visual comprehension based on Naive discriminative learning. Psychol. Rev. **118**(3), 438 (2011)

Balota, D.A., Yap, M.J., Hutchison, K.A., Cortese, M.J., Kessler, B., Loftis, B., et al.: The English lexicon project. Behav. Res. Methods **39**(3), 445–459 (2007)

Christiansen, M.H., Chater, N.: The now-or-never bottleneck: a fundamental constraint on language. Behav. Brain Sci. **39** (2016)

Elman, J.L.: Finding structure in time. Cogn. Sci. **14**(2), 179–211 (1990)

Garside, R.: A hybrid grammatical tagger: Claws 4, Corpus annotation. Linguistic information from computer text corpora (1997)

Hinojosa, J.A., Moreno, E.M., Casado, P., Muñoz, F., Pozo, M.A.: Syntactic expectancy: an event-related potentials study. Neurosci. Lett. **378**(1), 34–39 (2005)

Kingma, D.P., Ba, J.: Adam: a method for stochastic optimization (2014). arXiv preprint arXiv:1412.6980

Loper, E., Bird, S.: NLTK: the natural language toolkit. arXiv preprint cs/0205028 (2002)

Luke, S.G., Christianson, K.: Limits on lexical prediction during reading. Cogn. Psychol. **88**, 22–60 (2016)

Luke, S.G., Christianson, K.: The Provo corpus: a large eye-tracking corpus with predictability norms. Behav. Res. Methods **50**, 826–833 (2018)

Marcus, M., Santorini, B., Marcinkiewicz, M.A.: Building a large annotated corpus of English: The Penn Treebank (1993)

McClelland, J.L., Rumelhart, D.E.: An interactive activation model of context effects in letter perception: I. An account of basic findings. Psychol. Rev. **88**(5), 375 (1981)

McDonald, S.A., Shillcock, R.C.: Rethinking the word frequency effect: the neglected role of distributional information in lexical processing. Lang. Speech **44**(3), 295–322 (2001)

Mirman, D., Graf Estes, K., Magnuson, J.S.: Computational modeling of statistical learning: effects of transitional probability versus frequency and links to word learning. Infancy **15**(5), 471–486 (2010)

Misyak, J.B., Christiansen, M.H., Bruce Tomblin, J.: Sequential expectations: the role of prediction-based learning in language. Topics Cogn. Sci. **2**(1), 138–153 (2010)

Nivre, J., Abrams, M., Agić, Ž., et al.: Universal dependencies 2.3. (LINDAT/CLARIN digital library at the Institute of Formal and Applied Linguistics (ÚFAL), Faculty of Mathematics and Physics, Charles University) (2018)

Rescorla, R.A., Wagner, A.R., et al.: A theory of Pavlovian conditioning: variations in the effectiveness of reinforcement and nonreinforcement. Class. Conditioning II Curr. Res. Theory **2**, 64–99 (1972)

Srivastava, N., Hinton, G., Krizhevsky, A., Sutskever, I., Salakhutdinov, R.: Dropout: a simple way to prevent neural networks from overfitting. J. Mach. Learn. Res. **15**(1), 1929–1958 (2014)

Taylor, W.L.: "Cloze procedure": a new tool for measuring readability. J. Bull. **30**(4), 415–433 (1953)

Van Berkum, J.J., Brown, C.M., Zwitserlood, P., Kooijman, V., Hagoort, P.: Anticipating upcoming words in discourse: evidence from ERPs and reading times. J. Exp. Psychol. Learn. Mem. Cognit. **31**(3), 443 (2005)

Latent Space Exploration Using Generative Kernel PCA

David Winant[(✉)], Joachim Schreurs, and Johan A. K. Suykens

Department of Electrical Engineering (ESAT), STADIUS Center for Dynamical Systems, Signal Processing and Data Analytics, KU Leuven, Kasteelpark Arenberg 10, 3001 Leuven, Belgium
{david.winant,joachim.schreurs,johan.suykens}@kuleuven.be

Abstract. Kernel PCA is a powerful feature extractor which recently has seen a reformulation in the context of Restricted Kernel Machines (RKMs). These RKMs allow for a representation of kernel PCA in terms of hidden and visible units similar to Restricted Boltzmann Machines. This connection has led to insights on how to use kernel PCA in a generative procedure, called generative kernel PCA. In this paper, the use of generative kernel PCA for exploring latent spaces of datasets is investigated. New points can be generated by gradually moving in the latent space, which allows for an interpretation of the components. Firstly, examples of this feature space exploration on three datasets are shown with one of them leading to an interpretable representation of ECG signals. Afterwards, the use of the tool in combination with novelty detection is shown, where the latent space around novel patterns in the data is explored. This helps in the interpretation of why certain points are considered as novel.

Keywords: Kernel PCA · Restricted Kernel Machines · Latent space exploration

1 Introduction

Latent spaces provide a representation of data by embedding the data into an underlying vector space. Exploring these spaces allows for deeper insights in the structure of the data distribution, as well as understanding relationships between data points. Latent spaces are used for various purposes like latent space cartography [11], object shape generation [21] or style-based generation [8]. In this paper, the focus will be on how the synthesis of new data with generative methods can help with understanding the latent features extracted from a dataset. In recent years, generative methods have become a hot research topic within the field of machine learning. Two of the most well-known examples include variational autoencoders (VAEs) [9] and Generative Adversarial Networks (GANs) [2]. An example of a real-world application of latent spaces using VAEs is shown in [20], where deep convolutional VAEs are used to extract a biologically meaningful latent space from a cancer transcriptomes dataset. This latent space is used

© Springer Nature Switzerland AG 2020
B. Bogaerts et al. (Eds.): BNAIC 2019/BENELEARN 2019, CCIS 1196, pp. 70–82, 2020.
https://doi.org/10.1007/978-3-030-65154-1_5

to explore hypothetical gene expression profiles of tumors and their reaction to possible treatments. Similarly disentangled variational autoencoders have been used to find an interpretable and explainable representation of ECG signals [19]. Latent space exploration is also used for interpreting GANs, where interpolation between different images allows for the interpretation of the different features captured by the latent space, such as windows and curtains when working with datasets of bedroom images [14]. Latent space models are especially appealing for the synthesis of plausible pseudo-data with certain desirable properties. If the latent space is disentangled or uncorrelated, it is easier to interpret the meaning of different components in the latent space. Therefore it is easier to generate examples with desired properties, e.g. we want to generate a new face with certain characteristics. More recently, the concept of latent space exploration with GANs has been further developed by introducing new couplings of the latent space to the architecture of the generative network, this allows for control of local features for image synthesis at different scales in a style-based design [8]. These adaptations of GANs are known as Style-GANs. When applied to a facial dataset, the features can range from general face shape and hair style up to eyes, hair colour and mouth shape.

In this paper, kernel PCA is used as a generative mechanism [16]. Kernel PCA, as first described in [15], is a well-known feature extractor method often used for denoising and dimensionality reduction of datasets. Through the use of a kernel function it is a nonlinear extension to regular PCA by introducing an implicit, high dimensional latent feature space wherein the principal components are extracted. In [18], kernel PCA was cast within the framework of Restricted Kernel Machines (RKMs) which allows for an interpretation in terms of hidden and visible units similar to a type of generative neural network known as Restricted Boltzmann Machines (RBMs) [3]. This connection between kernel PCA and RBMs was later used to explore a generative mechanism for the kernel PCA [16]. A tensor-based multi-view classification model was introduced in [7]. In [13], a multi-view generative model called Generative RKM (Gen-RKM) is proposed which uses explicit feature-maps in a novel training procedure for joint feature-selection and subspace learning.

The goal of this paper is to explore the latent feature space extracted by kernel PCA using a generative mechanism, in an effort to interpret the components. This has led to the development of a Matlab tool which can be used to visualise the latent space of the kernel PCA method along its principal components. The use of the tool is demonstrated on three different datasets: the MNIST digits dataset, the Yale Face database and the MIT-BIH Arrhythmia database. As a final illustration, feature space exploration is used in the context of novelty detection [5], where the latent space around novel patterns in the data is explored. This to help the interpretation of why certain points are considered as novel.

In Sect. 2, a brief review on generative kernel PCA is given. In Sect. 3, latent feature space exploration is demonstrated. Subsequently we will illustrate how latent feature space exploration can help in interpreting novelty detection in Sect. 4. The paper is concluded in Sect. 5.

2 Kernel PCA in the RKM Framework

In this section, a short review on how kernel PCA can be used to generate new data is given, as introduced in [16]. We start with the calculation of the kernel principal components for a d-dimensional dataset $\{x_i\}_{i=1}^N$ with N data points and for each data point $x_i \in \mathbb{R}^d$. Compared to regular PCA, kernel PCA first maps the input data to a high dimensional feature space $\mathcal{F}$ using a feature map $\phi(\cdot)$. In this feature space, regular PCA is performed on the points $\phi(x_i)$ for $i = 1, \ldots, N$. By using a kernel function $k(x, y)$ defined as the inner product $(\phi(x) \cdot \phi(y))$, an explicit expression for $\phi(\cdot)$ can be avoided. Typical examples of such kernels are given by the Gaussian RBF kernel $k(x, y) = e^{-\|x-y\|_2^2/(2\sigma^2)}$ or the Laplace kernel $k(x, y) = e^{-\|x-y\|_2/\sigma}$, where σ denotes the bandwidth. Finding the principal components amounts to solving the eigenvalue problem for the kernel matrix[1] K, with matrix elements $K_{ij} = (\phi(\boldsymbol{x}_i) \cdot \phi(\boldsymbol{x}_j))$. The eigenvalue problem for kernel PCA is stated as follows:

$$KH^\top = H^\top \Lambda, \tag{1}$$

where $H = [h_1, \ldots, h_N] \in \mathbb{R}^{d \times N}$, the first $d \leq N$ components are used, is the matrix with the eigenvectors in each column and $\Lambda = \text{diag}\{\lambda_1, \ldots, \lambda_d\}$ the matrix with the corresponding eigenvalues on the diagonal. In the framework of RKMs, the points $\phi(x_i)$ correspond to visible units v_i and h_i are the hidden units. As in [16], the generative equation is given by:

$$v^\star = \phi(x^\star) = \left(\sum_{i=1}^N \phi(x_i) h_i^\top \right) h^\star, \tag{2}$$

where $h^\star$ represents a newly generated hidden unit and $v^\star$ the corresponding visible unit. Finding $x^\star$ in Eq. (2) corresponds to the pre-image problem [6]. In [16], the authors give a possible solution by multiplying both sides with $\phi(x_k)$, which gives the output of the kernel function for the generated point in the input space $x^\star$ and the data point x_k:

$$\hat{k}(x_k, x^\star) = \sum_{i=1}^N k(x_k, x_i) h_i^\top h^\star. \tag{3}$$

The above equation can be seen as the similarity between the newly generated point $x^\star$ and x_k. This expression can be used in a kernel smoother approach to find an estimate $\hat{x}$ for the generated data point $x^\star$:

$$\hat{x} = \frac{\sum_{i=1}^S \tilde{k}(x_i, x^\star) x_i}{\sum_{i=1}^S \tilde{k}(x_i, x^\star)}, \tag{4}$$

---

[1] For simplicity, the mapped data are assumed to be centered in $\mathcal{F}$. Otherwise, we have to go through the same algebra using $\tilde{\phi}(x) := \phi(x) - \sum_{i=1}^N \phi(x_i)$. This is the same assumption as in [15].

where $\tilde{k}(x_i, x^\star)$ is the scaled similarity between 0 and 1 calculated in (3) and S the number of closest points based on the similarity $\tilde{k}(x_i, x^\star)$. Given a point in the latent space $h^\star$, we get an approximation for the corresponding point $\hat{x}$ in input space. This mechanism makes it possible to continuously explore the latent space.

3 Experiments

Our goal is to use generative kernel PCA to explore the latent space. Therefore a tool[2] is developed where generative kernel PCA can easily be utilised for new datasets. First kernel PCA is performed to find the hidden features of the dataset. After choosing an initial hidden unit as starting point, the values are varied for each component of the hidden unit to explore the latent space. The corresponding newly generated data point in the input space is estimated using the kernel smoother approach.

In the tool, a partial visualisation of the latent space projected onto two principal components is shown. We continuously vary the values of the components of the selected hidden unit. This allows the exploration of the extracted latent space by visualising the resulting variation in the input space. The ability to perform incremental variations aids interpretation of the meaning encoded in the latent space along a chosen direction. In Fig. 1, the interface of our tool is shown.

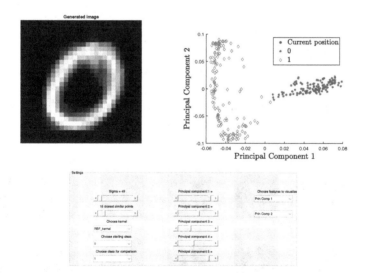

Fig. 1. Interface of the Matlab tool for exploring the latent space. At the bottom, the parameter values and position in the latent space can be chosen. In the top right the latent space along two selected principal components is shown and on the left the newly generated data point in the input space is visualised.

---

[2] Matlab code for the latent space exploration tool is available at https://www.esat. kuleuven.be/stadius/E/software.php.

MNIST Handwritten Digits

As an example, the latent space of the MNIST handwritten digits dataset [10] is explored, where 1000 data points each of digits zero and one are sampled. A Gaussian kernel with bandwidth $\sigma^2 = 50$, $S = 15$ and number of components $d = 10$. In Fig. 2, the latent space is shown along the first two principal components as well as the first and third components.

In Fig. 3, digits are generated along the directions indicated on the plots of the latent space in Fig. 2. This allows us to interpret the different regions and the meaning of the principal components. Along direction A, corresponding to the first principal component, we find an interpolation between the regions with digits of zero and one. Direction B seems to correlate with the orientation of the digit. This explains the smaller variation along the second principal component for the zeros as rotating the digit zero has a smaller effect compared to the rotation of digit one. The third direction, corresponding to component 3, seems to

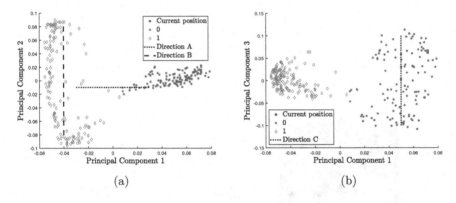

Fig. 2. Latent space of the MNIST digits dataset for the digits 0 and 1. The dotted lines indicate the direction along which new data points are generated. (a) Data projected on the first two principal components (b) Data projected on the first and third principal component.

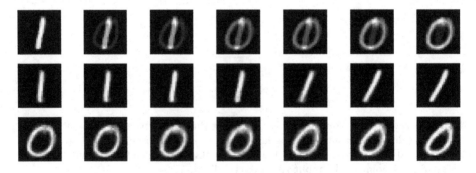

Fig. 3. Exploration of the latent space of the MNIST digits data set. In the top two rows images are generated along the directions A and B in Fig. 2a and in the bottom row the images are generated along direction C in Fig. 2b.

be related to squeezing the zeros together, which explains the larger variance for the zeros compared to the ones.

Yale Face Database

Another example of latent space exploration is done on the Extended Yale Face Database B [1], where 1720 data points are sampled. A Gaussian kernel with bandwidth $\sigma^2 = 650$, $S = 45$ and number of components $d = 20$.

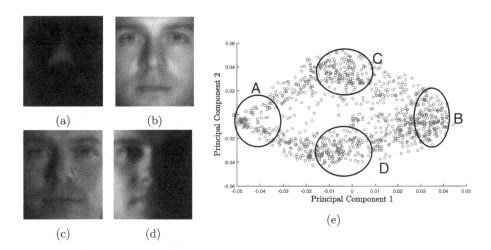

Fig. 4. Exploring different regions of the latent space of the Yale Face Database. (e) Data projected on the first two principal components for the Yale Face Database. (a)–(d) Generated faces from the different regions.

The latent space along the first two principal components is shown in Fig. 4e. Four different regions within the feature space are highlighted from which corresponding images are generated. The dissimilarity between the images in the various regions suggests the components capture different lighting conditions on the subjects.

The tool allows us to gradually move between these different regions and see the changes in the input space as shown in Fig. 6. Moving between regions A and B shows increasing illumination of the subject. We can thus interpret the first principal component as determining the global level of illumination. Note that besides data points without a light source no variation of the intensity of the lighting was varied while collecting the data for the Yale Face Database B. Only the position of the light source was changed. Generative Kernel PCA thus allows us to control the level of illumination regardless of the position of the light source. The bottom row seems to indicate that the second principal component can be interpreted as the position of the light source. In region C of the feature space the points are illuminated from the right and region D from the left. This interpretation of the second principal component seems indeed valid from Fig. 5a

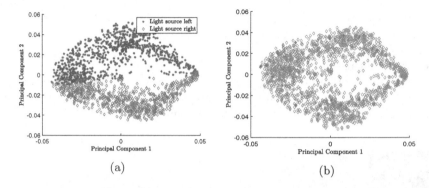

Fig. 5. Latent space of the Yale Facebase database B. (a) Points in orange indicate data points with a negative azimuthal angle between the camera direction and source of illumination, which corresponds to a light source to the right of the subject and vice versa for positive azimuthal angle. (b) Points in red indicate the hidden units of the same subject with lighting from different azimuthal angles. (Color figure online)

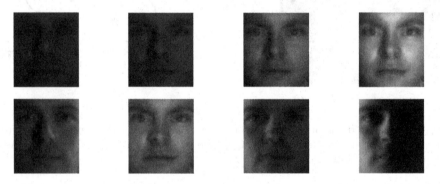

Fig. 6. Exploring the space between the regions of the latent space in Fig. 4e. The top row shows images generated between regions A and B, while the bottom row explores the space between regions C and D.

where the latent space is visualised with labels indicating the position of illumination obtained from the Yale Face Database. Faces with a positive azimuthal angle between the camera direction and the source of illumination are contained in the top half of the figure. This corresponds to a light source left of the subject and vice versa for a negative azimuthal angle. The first and second component are thus disentangled as the level of illumination does not determine whether the light comes from the left or the right. Furthermore in Fig. 5b the hidden units corresponding to the same subject under different lighting conditions are shown. The elevation of the light source is kept constant at zero elevation, while the azimuthal angle is varied. We see from the plot that the points move not strictly along the second principal component but follow a more circular path. This indicates that varying the azimuthal angle correlates with both the first and second principal component, i.e. moving the light source more to the side also decreases the global illumination level as less light is able to illuminate the face. We conclude that while in the original data set the position of the light source and the level of illumination are correlated, kernel PCA allows us to disentangle these factors and vary them separately when generating new images.

As a further example of generative kernel PCA, interpolation between 2 faces is demonstrated. Kernel PCA is performed on a subset of the database consisting out of 130 facial images of two subjects, the hyperparameters are the same as above. Variation along the fourth principal component results in a smooth interpolation between the two subjects, shown in Fig. 7. We also include an example in the bottom row where the interpolation does not result in a smooth change between the subjects. This illustrates a major limitation of our method as generative kernel PCA predominantly detects global features such as lighting and has difficulty with smaller, local features such as eyes. This stems from the fact that generative kernel PCA relies on the input data to be highly correlated which in this example translates itself to the need of the faces to be aligned with each other.

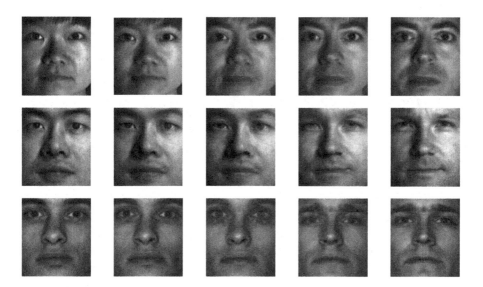

Fig. 7. Three examples of interpolation between two subjects of the Yale Face Database B along the fourth component. The uttermost left and right pictures in the rows represent the original faces.

MIT-BIH Arrhythmia Database

Besides the previous examples of latent space exploration for image datasets, kernel PCA is also applicable to other types of data. In this section, the MIT-BIH Arrhythmia dataset [12] is considered consisting out of ECG signals. The goal is to demonstrate the use of kernel PCA to extract interpretable directions in the latent feature space of the ECG signals. This would allow a clinical expert to gain insight and trust in the features extracted by the model. Similar research was previously done by [19] where they investigated the use of disentangled variational autoencoders to extract interpretable ECG embeddings. A similar approach is used to preprocess the data as in [19].

The signals from the patients with identifiers 101, 106, 103 and 105 are used for the normal beat signals and the data of patients 102, 104, 107, 217 for the paced beat signals. This results in a total of 785 beat patterns which are processed through a peak detection program [17]. The ECG signal is first passed through a fifth-order Butterworth bandpass filter with a lower cutoff frequency 1 Hz and upper cutoff frequency 60 Hz. The ECG beats are sampled 360 Hz and a window of 0.5 s is taken around each R-wave resulting in 180 samples per epoch. A regular Gaussian kernel with bandwidth $\sigma^2 = 10$ is used, with $S = 10$. The first 10 principal components are used in the reconstruction.

In Fig. 8 the latent feature space projected on the first two principal components is shown. Kernel PCA is also able to separate between the normal and paced beats.

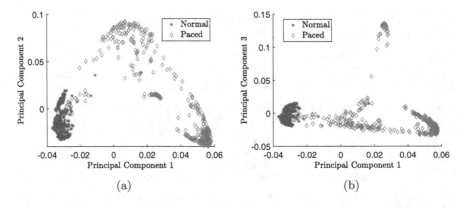

Fig. 8. The latent space for 785 ECG beat signals of the MIT-BIH Arrhythmia dataset projected on different principal components. The hidden units of both normal and paced heartbeats are shown.

Figure 9 shows the result in input space of moving along the first principal components in the latent feature space. As original base point we take a normal beat signal, i.e. corresponding to a hidden unit on the bottom right of Fig. 8a. The smooth transition between the beat patterns allows for interpretation of the first principal components. This allows a clinical expert to understand on what basis the paced beats are separated by the principal components and if this basis has a physiological meaning. In order to investigate the separated region of the latent space at the top of Fig. 8b we start from a paced beat pattern and vary along the third principal component. This allows us to see which sort of heartbeat patterns are responsible for this specific distribution in the latent space.

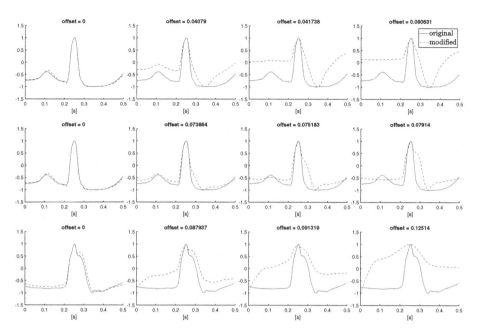

Fig. 9. Exploring the first three principal components of the latent feature space for the MIT-BIH arrhythmia database for normal and paced beats. The red line represents the newly generated datapoint compared to the original point depicted in blue. The top, middle and bottom row represent the variation along the first, second and third components respectively. Top and middle row start with a normal heartbeat pattern and the bottom row with a paced signal. (Color figure online)

4   Novelty Detection

As a final illustration of latent space exploration using generative kernel PCA, we consider an application within the context of novelty detection. We use the reconstruction error in feature space as a measure of novelty [4], where Hoffmann shows the metric demonstrates a competitive performance on synthetic distributions and real-world data sets. The novelty score is calculated for all data points, where the 20% of data points with the largest novelty score are considered novel. These points typically reside in low density regions of the latent space and are highlighted as interesting regions to explore using the tool. We consider 1000 instances of the digit zero from the MNIST dataset. After performing kernel PCA with the same parameters as in the previous section, we explore the latent space around the detected novel patterns. The data projected on the first two principal components is shown in Fig. 10.

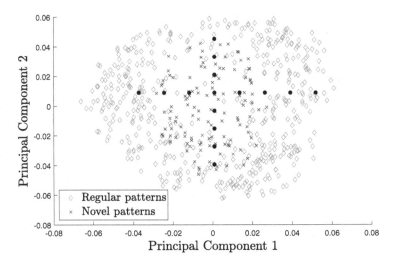

Fig. 10. The latent space for 1000 zeros of the MNIST digits data set. The central cluster of points consists out of data points with a high novelty score, this corresponds to a low density region in the latent space. The black dots indicate the points in latent space which are sampled.

The generated images from the positions indicated by the black dots in Fig. 10 are shown in Fig. 11. The first row allows us to interpret the first principal component as moving from a thin round zero towards a more closed digit. The middle of the latent space is where the novel patterns are located which seems to indicate most zeros are either thin and wide or thick and narrow. A low amount of zeros in the data set are thick and wide or very thin and narrow. The bottom row of Fig. 11 gives the interpretation for the second principal component as rotating the digit. The novel patterns seem to be clustered more together and as such have a less obvious orientation. Important to note is that we only look at the first 2 components for the interpretation, while in practice the novelty detection method takes all 20 components into consideration.

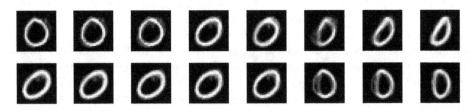

Fig. 11. Exploration of the latent space in Fig. 10. The top row indicates the points generated from the horizontal black dots, while the bottom row correspond to the vertical positions.

Above experiment shows that latent space exploration methods can give additional insights for novelty detection. Both the generating mechanism, as well as the novelty detection make use of the kernel PCA formulation. The two methods naturally complement each other: the novelty detection provides interesting regions in the latent space to explore, at the same time helps the generative mechanism in interpreting why certain points are considered as novel.

5 Conclusion

The use of generative kernel PCA in exploring the latent space is demonstrated. Gradually moving along components in the feature space allows for the interpretation of components and consequently additional insight into the underlying latent space. This mechanism is demonstrated on the MNIST handwritten digits data set, the Yale Face Database B and the MIT-BIH Arrhythmia database. The last example showed generative kernel PCA to be a interesting method for obtaining an interpretable representation of the ECG beat embedding. As a final illustration, feature space exploration is used in the context of novelty detection [5], where the latent space around novel patterns in data is explored. This to aid the interpretation of why certain points are considered as novel. Possible future directions would be the consideration of the geometry of the latent space. Not moving in straight lines, but curves through high density regions. Another direction would be to make use of different types of kernels as well as explicit feature maps for more flexibility in the latent feature space.

Acknowledgements. EU: The research leading to these results has received funding from the European Research Council under the European Union's Horizon 2020 research and innovation program/ERC Advanced Grant E-DUALITY (787960). This paper reflects only the authors' views and the Union is not liable for any use that may be made of the contained information. Research Council KUL: Optimization frameworks for deep kernel machines C14/18/068. Flemish Government: FWO: projects: GOA4917N (Deep Restricted Kernel Machines: Methods and Foundations), PhD/Postdoc grant. Flemish Government: This research received funding from the Flemish Government under the "Onderzoeksprogramma Artificiële Intelligentie (AI) Vlaanderen" programme.

References

1. Georghiades, A., Belhumeur, P., Kriegman, D.: From few to many: illumination cone models for face recognition under variable lighting and pose. IEEE Trans. Pattern Anal. Mach. Intell. **23**(6), 643–660 (2001)
2. Goodfellow, I., et al.: Generative adversarial nets. In: Advances in Neural Information Processing Systems, pp. 2672–2680 (2014)
3. Hinton, G.E.: A practical guide to training restricted boltzmann machines. In: Montavon, G., Orr, G.B., Müller, K.-R. (eds.) Neural Networks: Tricks of the Trade. LNCS, vol. 7700, pp. 599–619. Springer, Heidelberg (2012). https://doi.org/10.1007/978-3-642-35289-8_32

4. Hoffmann, H.: Kernel PCA for novelty detection. Pattern Recogn. **40**(3), 863–874 (2007)
5. Hofmann, T., Schölkopf, B., Smola, A.J.: Kernel methods in machine learning. The annals of statistics, pp. 1171–1220 (2008)
6. Honeine, P., Richard, C.: Preimage problem in kernel-based machine learning. IEEE Signal Process. Mag. **28**(2), 77–88 (2011)
7. Houthuys, L., Suykens, J.A.K.: Tensor Learning in multi-view kernel PCA. In: Kůrková, V., Manolopoulos, Y., Hammer, B., Iliadis, L., Maglogiannis, I. (eds.) ICANN 2018. LNCS, vol. 11140, pp. 205–215. Springer, Cham (2018). https://doi.org/10.1007/978-3-030-01421-6_21
8. Karras, T., Laine, S., Aila, T.: A style-based generator architecture for generative adversarial networks. In: Proceedings of the IEEE Conference on Computer Vision and Pattern Recognition, pp. 4401–4410 (2019)
9. Kingma, D.P., Welling, M.: Auto-encoding variational Bayes. arXiv preprint arXiv:1312.6114 (2013)
10. LeCun, Y., Bottou, L., Bengio, Y., Haffner, P., et al.: Gradient-based learning applied to document recognition. Proc. IEEE **86**(11), 2278–2324 (1998)
11. Liu, Y., Jun, E., Li, Q., Heer, J.: Latent space cartography: visual analysis of vector space embeddings. In: Computer Graphics Forum, vol. 38, pp. 67–78. Wiley Online Library (2019)
12. Moody, G.B., Mark, R.G.: The impact of the MIT-BIH arrhythmia database. IEEE Eng. Med. Biol. Mag. **20**(3), 45–50 (2001)
13. Pandey, A., Schreurs, J., Suykens, J.A.K.: Generative restricted kernel machines. arXiv preprint arXiv:1906.08144 (2019)
14. Radford, A., Metz, L., Chintala, S.: Unsupervised representation learning with deep convolutional generative adversarial networks. arXiv preprint arXiv:1511.06434 (2015)
15. Schölkopf, Bernhard., Smola, Alexander., Müller, Klaus-Robert: Kernel principal component analysis. In: Gerstner, Wulfram, Germond, Alain, Hasler, Martin, Nicoud, Jean-Daniel (eds.) ICANN 1997. LNCS, vol. 1327, pp. 583–588. Springer, Heidelberg (1997). https://doi.org/10.1007/BFb0020217
16. Schreurs, J., Suykens, J.A.K.: Generative kernel PCA. ESANN **2018**, 129–134 (2018)
17. Sedghamiz, H.: An online algorithm for r, s and t wave detection. Linkoping University, December 2013
18. Suykens, J.A.K.: Deep restricted kernel machines using conjugate feature duality. Neural Comput. **29**(8), 2123–2163 (2017)
19. Van Steenkiste, T., Deschrijver, D., Dhaene, T.: Interpretable ECG beat embedding using disentangled variational auto-encoders. In: 2019 IEEE 32nd International Symposium on Computer-Based Medical Systems (CBMS), pp. 373–378. IEEE (2019)
20. Way, G.P., Greene, C.S.: Extracting a biologically relevant latent space from cancer transcriptomes with variational autoencoders, p. 174474. BioRxiv (2017)
21. Wu, J., Zhang, C., Xue, T., Freeman, B., Tenenbaum, J.: Learning a probabilistic latent space of object shapes via 3D generative-adversarial modeling. Adv. NeurIPS. **29**, 82–90 (2016)

Machine Learning Part: Benelearn

Calibrated Multi-probabilistic Prediction as a Defense Against Adversarial Attacks

Jonathan Peck[1,2]($\boxtimes$), Bart Goossens[3], and Yvan Saeys[1,2]

[1] Department of Applied Mathematics, Computer Science and Statistics,
Ghent University, 9000 Ghent, Belgium
{Jonathan.Peck,Yvan.Saeys}@ugent.be
[2] Data Mining and Modeling for Biomedicine, VIB Inflammation Research Center,
9052 Ghent, Belgium
[3] Department of Telecommunications and Information Processing,
IMEC/Ghent University, 9000 Ghent, Belgium
Bart.Goossens@ugent.be

Abstract. Machine learning (ML) classifiers—in particular deep neural networks—are surprisingly vulnerable to so-called *adversarial examples.* These are small modifications of natural inputs which drastically alter the output of the model even though no relevant features appear to have been modified. One explanation that has been offered for this phenomenon is the *calibration hypothesis*, which states that the probabilistic predictions of typical ML models are miscalibrated. As a result, classifiers can often be very confident in completely erroneous predictions. Based on this idea, we propose the *MultIVAP* algorithm for defending arbitrary ML models against adversarial examples. Our method is inspired by the *inductive Venn-ABERS predictor (IVAP)* technique from the field of conformal prediction. The IVAP enjoys the theoretical guarantee that its predictions will be *perfectly calibrated*, thus addressing the problem of miscalibration. Experimental results on five image classification tasks demonstrate empirically that the MultIVAP has a reasonably small computational overhead and provides significantly higher adversarial robustness without sacrificing accuracy on clean data. This increase in robustness is observed both against defense-oblivious attacks as well as a defense-aware white-box attack specifically designed for the MultIVAP.

We make our code available at https://github.com/saeyslab/multivap.

1 Introduction

Machine learning techniques have made great progress in recent years, obtaining state of the art performance in areas such as natural language processing [76] as well as image and speech recognition [73]. However, the theoretical properties of

We thank the NVIDIA Corporation for the donation of a Titan Xp GPU with which we were able to carry out our experiments. Jonathan Peck is sponsored by a fellowship of the Research Foundation Flanders (FWO). Yvan Saeys is an ISAC Marylou Ingram scholar.

© Springer Nature Switzerland AG 2020
B. Bogaerts et al. (Eds.): BNAIC 2019/BENELEARN 2019, CCIS 1196, pp. 85–125, 2020.
https://doi.org/10.1007/978-3-030-65154-1_6

ice cream   burrito

 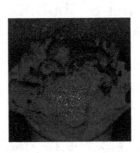 

Fig. 1. Example of an adversarial perturbation for image classification models. The image on the left is correctly classified as ice cream by the ResNet50 deep neural network from [36]. When the perturbation shown in the middle is added to this image, we obtain the sample on the right. Although visually indistinguishable, the altered image is classified as a burrito instead.

the deep neural networks responsible for this success remain poorly understood. At present, there is no theory which can satisfactorily explain the success of deep learning and many open questions remain [91]. A peculiar example of this lack of theoretical understanding is the existence of so-called *adversarial perturbations* [6]. These are small modifications to the inputs of a model which can drastically change its output, even though the alterations are completely insignificant. This is perhaps nowhere as apparent as in image recognition, which is where the phenomenon was first studied for deep neural networks [80]. Figure 1 shows an example of an adversarial perturbation in this setting. The phenomenon is by no means restricted to images, however: recently, significant progress has been made towards generating adversarial examples for domains other than images. This includes speech recognition, where inaudible distortions of spoken fragments can lead to erroneous transcriptions [65], and text processing where small typographical errors can mislead the models [22].

As pointed out by [6], adversarial examples already have a long history. However, to date, it appears there is no consensus regarding the causes of this vulnerability in modern deep neural networks nor do there appear to exist satisfactory solutions. Despite intense research effort, at present there does not exist any real defense against this phenomenon; almost all serious defenses which have been proposed thus far have either been broken completely [4,10,12,13] or have difficulties scaling to realistic problems [16,51,67]. This is problematic for a few reasons:

- From a statistical learning perspective, it calls into question how well our ML models actually grasp the task they are expected to solve. If the accuracy of an image recognition algorithm can be degraded to the level of random guessing (or even below that) by changing just a handful of pixels in a high-resolution image, then what has this model actually learned?
- There are several plausible scenarios where malicious actors may be both capable as well as tempted to exploit these shortcomings for personal gain [29].

Examples of this include bypassing automated content filtering systems and facial recognition scanners, detecting copyright violations, flagging e-mail spam, etc.
- Several powerful attacks exist which can successfully fool ML models using adversarial examples even when little to no knowledge about the target model is available [8,14,35]. These so-called "black-box attacks" usually only require query access to the model and can generate highly effective adversarials with a limited number of queries. However, some methods do not even require query access: *universal adversarial perturbations* can fool a wide range of ML models and can be crafted without any form of access to the target model [55].

One possible explanation for this problem that has garnered some support is the *calibration hypothesis* [19,34]. This idea explains the existence of adversarial examples as a consequence of overconfident extrapolation by current machine learning models. More formally, the calibration hypothesis states that the predictions of our models are not *calibrated*. To understand precisely what this means, consider a typical parametric machine learning model for multiclass classification based on a finite set of parameters $\theta \in \Theta$. Such a model often works by fitting a *scoring function* $s : \mathcal{X} \times \Theta \to \mathbb{R}^k$ which yields a vector of real numbers, one component per class. These scores are used to estimate probabilities $p_i(x; \theta) \approx \Pr[Y = i \mid X = x]$—that is, the probability that a given sample x belongs to class i—via some form of *calibration*. The most common method is Platt's scaling [63], which fits a logistic sigmoid to the score vector:

$$p_i(x; \theta) = \frac{\exp(w_i s_i(x; \theta) + b_i)}{\sum_{j=1}^{k} \exp(w_j s_j(x; \theta) + b_j)}.$$

The parameters θ of the model are then chosen so that the Kullback-Leibler divergence between the data distribution and the model distribution is minimized. In practice, this corresponds to minimizing the negative log-likelihood of the data $(x_1, y_1), \ldots, (x_m, y_m)$ under the model distribution:

$$\theta^\star = \arg\min_\theta \; - \sum_{i=1}^{m} \log p_{y_i}(x_i; \theta). \tag{1}$$

The final predictions are then given by $\hat{y}(x) = \arg\max_i \; p_i(x; \theta^\star)$. These are *point predictions* in the sense that only a single output $\hat{y}(x) \in \mathcal{Y}$ is given. If any measure of uncertainty accompanies these predictions at all, it is usually the estimated class probability $p_i(x; \theta^\star)$. However, as discussed in [19], [86] and [34], these probabilities can be badly mistaken: often there is no theoretical guarantee that the estimated probability $p_i(x)$ will approximate the true probability $\Pr[Y = i \mid X = x]$ to any degree of accuracy[1]. In particular, [34] perform an

---

[1] Of course, there is the classic result of [25] which states that, in expectation, the cross entropy loss is minimized if and only if the model perfectly recovers the data distribution. In practice, however, we rarely minimize this loss exactly. It is currently a major open problem in deep learning to provide similar guarantees when the model fit is suboptimal.

extensive comparison of modern deep neural networks with older methods. They come to the surprising conclusion that miscalibration is especially pronounced among *newer* models, even though these models achieve superior classification performance to the older methods.

Note that this phenomenon is perfectly consistent with the existence of efficient learning algorithms which output highly accurate hypothesis functions, because a low classification error only implies that the class that is assigned the highest probability will usually also be the correct class. There is almost never any penalty on *overconfidence*: if the true class only has a probability of 60%, say, then the prediction will be counted as correct even if the learned classifier assigns it a probability of 100%. Indeed, one almost never has access to such precise probability estimates; in practice, data sets used for supervised learning only contain the most likely label without any accompanying probabilities. The typical trick to remedy this is to assume the most likely label has probability one and fit the classifier accordingly. In some sense, this practice in fact encourages overconfidence and so the issue of adversarial vulnerability due to miscalibration may come as no surprise. A popular method to address this flaw is *label smoothing* [79]. This technique linearly interpolates between the probability distribution produced by the model and another distribution, typically the uniform distribution, in order to "soften" the predictions and introduce more uncertainty:

$$p'_i(x; \theta) = (1 - \varepsilon)p_i(x; \theta) + \varepsilon u(i).$$

If u is the uniform distribution on k classes, this reduces to

$$p'_i(x; \theta) = (1 - \varepsilon)p_i(x; \theta) + \frac{\varepsilon}{k}.$$

Label smoothing has been used with some success in adversarial settings [44], but it lacks any theoretical guarantees.

Although the calibration hypothesis has not yet been proven by any means, some reasonable arguments exist that make it a plausible candidate for explaining at least in part why classifiers can be so brittle. For one, the maximum likelihood estimate θ^* returned by the optimization (1) can incur approximation and estimation errors due to biases in the data set and ill-specified model classes. Often, in order to solve (1) in practice, optimization algorithms such as gradient descent are used which are scalable but lack theoretical convergence guarantees if the problem is not convex[2] [7,30]. Hence, we might not achieve a global optimum or even a good local optimum of the cost function. Moreover, it is difficult to quantify exactly how reliable any given point prediction made by a deep neural network really is [26,54]. ReLU networks in particular can severely overestimate class membership probabilities when data points lie far away from previously seen training data [37]. It therefore makes sense that an adversary

---

[2] Guarantees can also be given for non-convex problems, but these usually require at least bounded iterates or a Lipschitz continuous gradient [78]. Such assumptions are often violated or difficult to verify in practice.

could exploit all of these deficiencies to deliberately craft input samples on which the model will be overconfident in the wrong direction.

If one accepts the calibration hypothesis, then it is natural to look into machine learning methods that enjoy provable guarantees on the calibration of their predictions. Such methods already exist: they are the main focus of the field of *conformal prediction* [72,85]. In this work, we employ one algorithm from this field in particular: the *inductive Venn-ABERS predictor (IVAP)* proposed by [86]. This is a computationally efficient technique which can hedge the predictions of any *binary* classifier in such a way that one obtains perfectly calibrated outputs. Our method, which we call the *MultIVAP*, extends the IVAP to multiclass problems and is able to provide significantly increased adversarial robustness. Empirical evaluations on five image classification tasks as well as theoretical results also support the idea that our defense can be scaled to realistic models.

1.1 Related Work

Research interest in adversarial robustness has increased dramatically since 2013, when [80] showed that deep neural networks can be very sensitive to small perturbations of their inputs. However, the idea of adversarial classification goes back to at least 2004 with the work of [18] who noticed that machine learning algorithms are highly sensitive to violations of the assumption that the data are independent and identically distributed. They proposed a solution for the case of naive Bayes classifiers. [49] continued this line of research and defined the *adversarial classifier reverse engineering (ACRE)* learning problem, where one has to learn sufficient information about the target machine learning system in order to construct adversarial attacks. They present efficient learning algorithms which reverse engineer linear classifiers and construct effective adversarial attacks. This idea was later generalized by [84], who proved that any distribution which can be efficiently learned can also be efficiently reverse engineered. For a historical overview of this field, we refer the reader to [6].

Since the work of [80], the ML community has focused mostly on adversarial robustness for deep neural networks. Important contributions in this vein include [31], who proposed the *linearity hypothesis* which explains the existence of adversarial examples as a consequence of overly linear behavior of deep neural networks. They also suggested the technique of *adversarial training*, a form of data augmentation which is still considered to be one of the most effective defenses. Subsequent work has challenged the linearity hypothesis [81] and the consensus at present seems to be that this idea indeed does not capture the full problem. Current attempts at defending against adversarial perturbations usually cast it as a robust optimization problem [51] and try to find certifiable lower bounds on the robustness achieved by the defense [16,67,74]. Some progress has also been made on theoretical questions of learnability and sample efficiency of robust classifiers [3,17,32,69,89]. The results of these investigations are mostly negative, indicating that robust learning is fundamentally harder and requires more data than standard learning.

To the best of our knowledge, aside from our own work in [62], little prior work has considered using conformal prediction as a defense against adversarial attacks. [61] propose a deep k-nearest neighbor classification scheme inspired by the conformal prediction algorithm from [85]. Not many evaluations of the robustness of this method appear to have been performed thus far, however: we are only aware of the work of [75] in this regard, who develop a moderately strong attack against the scheme. Moreover, this method requires a k-nearest neighbor search in high-dimensional spaces, so its computational efficiency is questionable. [9] augment existing classifiers with a novel non-conformity measure in order to improve robustness to out-of-sample data, but they do not specifically evaluate this method against adversarial attacks and so its robustness against worst-case manipulations is unclear. Moreover, robustness to out-of-sample data may not be sufficient for general adversarial robustness: some attacks explicitly try to generate adversarials that lie on the data manifold. These cannot be considered as out-of-distribution and yet they still very often succeed in fooling models [24, 43].

We cannot discuss calibration of ML model predictions without also mentioning *Bayesian neural networks* or BNNs [26]. In the Bayesian setting, one does not consider the optimal point estimate θ^* of the model parameters; rather, one computes a weighted average over *all possible models*. This requires the specification of a prior $\Pr[\theta]$ on the model weights. Inference in a BNN then involves marginalization over these weights:

$$\Pr[y \mid x] = \int \Pr[\theta] \Pr[y \mid x, \theta] \mathrm{d}\theta.$$

The exact computation of this integral is intractable for realistic problems where θ is high-dimensional and $\Pr[y \mid x, \theta]$ is given by a complicated deep neural network. As such, various approximation techiques such as sampling, variational inference, Monte Carlo Dropout and SWAG have been proposed [27,42,50]. These methods have faced criticism, however, mostly because there is disagreement as to how one should specify the prior. It is usually argued that *uninformative* distributions such as the standard Jeffreys prior can be used for this purpose [41]. On the contrary, [28] argue the exact opposite: Bayesian neural networks require *highly informative* priors, otherwise they cannot provide useful uncertainties. The main issue is that neural networks are often *overparameterized*: they have many more parameters than training data points to fit them. Combined with an uninformative prior, this leads to a posterior distribution $\Pr[y \mid x]$ that does not *concentrate*, that is, which cannot distinguish predictors that generalize well from ones that do not. Hence, BNNs run the risk of not yielding useful uncertainty estimates if an insufficiently informative prior is used.

By contrast, the method we investigate here is a frequentist procedure, relying on the computation of p-values derived from some measure of "conformity". This conformity score does not require us to form uncertainty estimates of the weights of the model; instead, it is calculated based solely on the scores assigned by the classifier to samples from a separate *calibration data set*. As a result, this method has much smaller computational overhead than the Bayesian techniques, but it requires us to sacrifice part of the data set for calibration.

It is interesting to note, however, that even though our approach is frequentist, it can easily be combined with Bayesian techniques since it makes no assumptions on the underlying model. In particular, our algorithm can be applied to a BNN in order to improve its calibration. This combination of Bayesian inference with frequentist methods has been called *Frasian inference* [87], a portmanteau of *frequentist* and *Bayesian* as well as a tribute to the statistician Donald Fraser. Whether such combinations of frequentist and Bayesian procedures have any significant merit for adversarial robustness is an interesting question that we leave for future work.

We also mention the work of [52] who consider another generalization of the IVAP to multiclass problems. Their approach is based on a one-vs-one classification scheme where a separate IVAP is used for each pair of distinct classes i and j. They then estimate the pairwise class probabilities

$$r_{ij}(x) \approx \Pr[Y = i \mid Y \in \{i, j\}, X = x].$$

In the case of an IVAP, which returns pairs of probabilities (p_0^{ij}, p_1^{ij}), there are various ways one could reduce the output to a single probability estimate. A typical one is the minimax method:

$$r_{ij}(x) = \frac{p_1^{ij}(x)}{1 - p_0^{ij}(x) + p_1^{ij}(x)}.$$

These estimates are then merged into multiclass probabilities using a voting method due to [64]:

$$p_i(x) = \left(\sum_{j \neq i} \frac{1}{r_{ij}(x)} - (K - 2) \right)^{-1},$$

where K is the number of classes. [52] empirically study the performance of multiclass classifiers when their predictions are calibrated according to this strategy. They find that the above method using an IVAP is more reliable than existing popular alternatives such as Platt's scaling. The method we propose here is similar, but ours is based on a *one-vs-all* approach instead of a one-vs-one scheme. Moreover, it is not immediately clear what theoretical statistical guarantees the method proposed by [52] retains. Since it is our goal to provide some certifiable method for defending against adversarial perturbations, we are more concerned with provable guarantees on the calibration of the output of the classifier. To this end, we construct our algorithm in such a manner that calibration guarantees can still be given. Specifically, the long-term error rate ε is a controllable parameter of our method, similar to other conformal predictors [85].

1.2   Organization

The rest of this paper is organized as follows. In Sect. 2, we establish the necessary background in conformal prediction. Section 3 introduces our proposed defense, which we have called the *MultIVAP* algorithm. Experiments with this algorithm are carried out in Sect. 4. Finally, Sect. 5 concludes the work.

2   Conformal Prediction

As its name suggests, the field of conformal prediction is concerned with making predictions based on how well a previously unseen sample "conforms" to the data that has already been seen [85]. A *conformal predictor* is a function Γ which takes a confidence level $\varepsilon \in [0,1]$, a bag[3] of instances $B = \{(x_1, y_1), \ldots, (x_n, y_n)\}$ and a new object $x \in \mathcal{X}$ and outputs a set $\Gamma^\varepsilon(B, x) \subseteq \mathcal{Y}$ of possible labels. Intuitively, the set $\Gamma^\varepsilon(B, x)$ consists of those labels which the predictor believes might be the true label with a probability of at least $1-\varepsilon$. This property is called *exact validity* [85]. Formally, exact validity can be defined as follows. Let $\omega = (x_1, y_1), (x_2, y_2), \ldots$ be an infinite sequence of samples from an exchangeable[4] distribution. Define the error after n samples at significance level ε as

$$\mathrm{err}_n^\varepsilon(\Gamma, \omega) = \begin{cases} 1 & \text{if } y_n \notin \Gamma^\varepsilon(\{(x_1, y_1), \ldots, (x_{n-1}, y_{n-1})\}, x_n), \\ 0 & \text{otherwise.} \end{cases}$$

Exact validity means that the random variables $\mathrm{err}_1^\varepsilon(\Gamma, \omega), \mathrm{err}_2^\varepsilon(\Gamma, \omega), \ldots$ are all independent and Bernoulli distributed with parameter ε. For *deterministic* predictors, however, this property can never be satisfied (see [85] for a proof of this statement). Therefore, in the deterministic case, we only demand that the errors are *dominated in distribution* by a sequence of independent Bernoulli distributed random variables. That is, there exist two sequences of random variables $\xi_1, \xi_2, \ldots$ and $\eta_1, \eta_2, \ldots$ such that the following conditions are all met:

1. Each ξ_n is independent and Bernoulli distributed with parameter ε.
2. $\mathrm{err}_n^\varepsilon(\Gamma, \omega) \leq \eta_n$ almost surely for all n.
3. The joint distribution of $\eta_1, \ldots, \eta_n$ equals that of $\xi_1, \ldots, \xi_n$ for all n.

This property is called *conservative validity* and implies the following asymptotic result by the law of large numbers:

$$\lim_{n \to \infty} \Pr\left[\frac{1}{n} \sum_{i=1}^n \mathrm{err}_i^\varepsilon(\Gamma, \omega) \leq \varepsilon\right] = 1.$$

Hence, for deterministic conformal predictors, one can still say that the fraction of mistakes they make tends to be bounded by ε as n grows large. We can quantify this convergence more precisely as follows. Let

$$E_n = \frac{1}{n} \sum_{i=1}^n \mathrm{err}_i^\varepsilon(\Gamma, \omega).$$

---

[3] A *bag* or *multiset* is a collection of objects where the order is irrelevant (like a set) but duplicates are allowed (like a list).

[4] A probability distribution is said to be *exchangeable* if every permutation of a sequence is equally likely.

The sequence of samples ω satisfies $\mathbb{E}[\|E_n\|] < \infty$. Therefore, the sequence of conditional expectations $Z_t = \mathbb{E}[E_n \mid \omega_1, \ldots, \omega_t]$ forms a *Doob martingale* [21]. Furthermore, this martingale is $(1/n)$-Lipschitz:

$$|Z_{t+1} - Z_t| \leq \frac{1}{n}.$$

The Azuma-Hoeffding inequality [5,38,53] now yields

$$\Pr[E_n \geq \mathbb{E}[E_n] + t] \leq \exp(-2nt^2).$$

By conservative validity, we have $\mathbb{E}[E_n] \leq \varepsilon$ and so

$$\Pr[E_n \geq \varepsilon + t] \leq \exp(-2nt^2).$$

Setting $\delta = \exp(-2nt^2)$ we find

$$t = \sqrt{\frac{1}{2n} \log \frac{1}{\delta}}.$$

Therefore, the following inequality holds with probability at least $1 - \delta$:

$$\frac{1}{n} \sum_{i=1}^{n} \mathrm{err}_i^\varepsilon(\Gamma, \omega) \leq \varepsilon + \sqrt{\frac{1}{2n} \log \frac{1}{\delta}}. \tag{2}$$

Indeed, for $n \to \infty$ we have $E_n \leq \varepsilon$ almost surely. However, (2) allows us to construct confidence intervals when n is finite without having to make any distributional assumptions. Note that (2) is non-vacuous only for

$$n \geq \frac{1}{2(1 - \varepsilon)^2} \log \frac{1}{\delta}, \qquad\qquad \delta \geq \exp\left(-2n(1 - \varepsilon)^2\right).$$

For example, if we wish to guarantee (2) with probability at least 95% for $\varepsilon = 0.01$ and $n = 10^4$, then the error term is 0.012. This means we have $E_n \leq 2.2\%$ with probability exceeding 95% for a dataset consisting of 10k samples, which is a reasonably small sample size by current standards. Therefore, although conservative validity is purely an asymptotic result, favorable finite-sample bounds can also be given via typical concentration of measure arguments.

Algorithm 1 is the general conformal prediction algorithm. It takes as a parameter a *non-conformity measure* Δ. A non-conformity measure is any measurable real-valued function which takes a sequence of samples $z_1, \ldots, z_n$ along with an additional sample z and maps them to a *non-conformity score* $\Delta(\langle z_1, \ldots, z_n \rangle, z)$. This score is intended to measure how much the sample z differs from the given bag of samples. For example, in a regression problem, we might define a non-conformity measure by taking the absolute or squared difference between the given output y and the output $\hat{y}$ which we would predict given the previously seen examples.

Algorithm 1: The conformal prediction algorithm.

Input: Non-conformity measure Δ, significance level ε, bag of examples
$\{z_1, \ldots, z_n\} \subseteq \mathcal{Z}$, object $x \in \mathcal{X}$
Output: Prediction region $\Gamma^\varepsilon(\{z_1, \ldots, z_n\}, x)$

1 **foreach** $y \in \mathcal{Y}$ **do**
2      $z_{n+1} \leftarrow (x, y)$
3   **for** $i = 1, \ldots, n+1$ **do**
4          $\alpha_i \leftarrow \Delta(\{z_1, \ldots, z_n\} \setminus \{z_i\}, z_i)$
5   **end**
6      $p_y \leftarrow \frac{1}{n+1} \# \{i = 1, \ldots, n+1 \mid \alpha_i \geq \alpha_{n+1}\}$
7 **end**
8 **return** $\{y \in \mathcal{Y} \mid p_y > \varepsilon\}$

The conformal prediction algorithm is always conservatively valid regardless of the choice of non-conformity measure. However, the *predictive efficiency* of the algorithm—that is, the size of the prediction region $\Gamma^\varepsilon(B, x)$—can vary considerably with different choices for Δ. If the non-conformity measure is chosen sufficiently poorly, the prediction regions may even be equal to the entirety of $\mathcal{Y}$. Although this is clearly valid, it is useless from a practical point of view.

Algorithm 1 determines a prediction region for a new input $x \in \mathcal{X}$ based on a bag of old samples by iterating over every label $y \in \mathcal{Y}$ and computing an associated p-value p_y. This value is the empirical fraction of samples in the bag (including the new "virtual sample" (x, y)) with a non-conformity score that is at least as large as the non-conformity score of (x, y). By thresholding these p-values we obtain a subset of candidate labels $y_{i_1}, \ldots, y_{i_t}$ such that each possible combination $(x, y_{i_1}), \ldots, (x, y_{i_t})$ is "sufficiently conformal" to the old samples at the given level of confidence. More concretely, this means each of the labels $y_{i_1}, \ldots, y_{i_t}$ could be assigned to the sample x based on our knowledge of previously seen data. Conformal predictors are thus inherently *multi-label* predictors as they return sets of possible outcomes instead of a single outcome.

2.1 Inductive Venn-ABERS Predictors

In this work, we will mainly be interested in a particular conservatively valid predictor known as the *inductive Venn-ABERS predictor* (IVAP). First proposed by [86], an IVAP works by taking advantage of another inductive learning rule (such as a neural network) and calibrating its output in order to hedge the predictions of that rule.

The IVAP is designed only for *binary* classification problems. When instantiated with a scoring rule and an additional calibration data set, it outputs two scalars $p_0, p_1 \in [0, 1]$ with $p_0 \leq p_1$ for each new prediction. The theoretical guarantee enjoyed by the IVAP is that these quantities form bounds on the conditional probability that the true label is 1 given the input:

$$p_0(x) \leq \Pr[Y = 1 \mid X = x] \leq p_1(x).$$

To achieve this, the IVAP requires a *calibration set*, a separate data set that must be independent of the training set. It uses this additional data to fit a calibration curve with which we can assess the reliability of any new prediction. Algorithm 2 shows the pseudo-code.

Algorithm 2: The inductive Venn-ABERS prediction algorithm.

   Input: bag of examples $\{z_1, \ldots, z_n\} \subseteq \mathcal{Z}$, object $x \in \mathcal{X}$, learning
            algorithm A
   Output: Pair of probabilities (p_0, p_1)
1 Divide the bag of training examples $\{z_1, \ldots, z_n\}$ into a proper training set
   $\{z_1, \ldots, z_m\}$ and a calibration set $\{z_{m+1}, \ldots, z_n\}$.
2 Run the learning algorithm A on the proper training set to obtain a
   scoring rule F.
3 **foreach** *example $z_i = (x_i, y_i)$ in the calibration set* **do**
4 $\quad\mid\quad s_i \leftarrow F(x_i)$
5 **end**
6 $s \leftarrow F(x)$
7 Fit isotonic regression to $\{(s_{m+1}, y_{m+1}), \ldots, (s_n, y_n), (s, 0)\}$ obtaining a
   function f_0.
8 Fit isotonic regression to $\{(s_{m+1}, y_{m+1}), \ldots, (s_n, y_n), (s, 1)\}$ obtaining a
   function f_1.
9 $(p_0, p_1) \leftarrow (f_0(s), f_1(s))$
10 **return** (p_0, p_1)

IVAPs are a variant of the conformal prediction algorithm where the non-conformity measure is based on an isotonic regression of the scores which the underlying scoring classifier assigns to the calibration data points as well as the new input to be classified. Isotonic (or monotonic) regression aims to fit a non-decreasing free-form line to a sequence of observations such that the line lies as close to these observations as possible. In the case of Algorithm 2, the isotonic regression is performed as follows. Let $s_1, \ldots, s_k$ be the scores assigned to the calibration points. First, these points are sorted in increasing order and duplicates are removed, obtaining a sequence $s'_1 \leq \cdots \leq s'_t$. We then define the *multiplicity* of s'_j as

$$w_j = \#\{i \mid s_i = s'_j\}.$$

The "average label" corresponding to some score s'_j is

$$y'_j = \frac{1}{w_j} \sum_{i:s_i=s'_j} y_i.$$

The *cumulative sum diagram* (CSD) is computed as the set of points

$$P_i = \left(\sum_{j=1}^{i} w_j, \sum_{j=1}^{i} y'_j w_j \right) = (W_i, Y_i)$$

for $i = 1, \ldots, t$. For these points, the *greatest convex minorant* (GCM) is computed. Figure 2 shows an example of a GCM computed for a given set of points in the plane. Formally, the GCM of a function $f : U \to \mathbb{R}$ is the maximal convex function $g : I \to \mathbb{R}$ defined on a closed interval I containing U such that $g(u) \leq f(u)$ for all $u \in U$ [83]. It can be thought of as the "lowest part" of the convex hull of the graph of f.

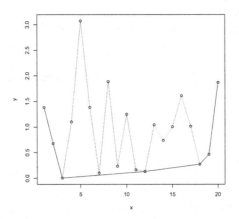

Fig. 2. Example of the greatest convex minorant of a set of points, shown as a red line. This image was produced with the `fdrtool` package [46].

The value at s'_i of the isotonic regression is now defined as the slope of the GCM between W_{i-1} and W_i. That is, if f is the isotonic regression and g is the GCM, then

$$f(s'_i) = \frac{g(W_i) - g(W_{i-1})}{W_i - W_{i-1}} = \frac{g(W_i) - g(W_{i-1})}{w_i}.$$

We leave the values of f at other points besides the s'_i undefined, as we will never need them here.

Figure 3 illustrates what the isotonic regression could look like for an IVAP. The scores s are shown as black circles. They were generated from a standard normal distribution and sorted in increasing order (as the IVAP does). The labels y were assigned based on whether the drawn sample was non-negative ($y = 1$ for $s \geq 0$) or negative ($y = 0$ for $s < 0$). Each label was flipped with some pre-determined probability of error that varies for each individual plot shown in the figure. The noise level for y increases from left to right and top to bottom, with the top-left plot being completely free of noise and the bottom-right plot consisting almost entirely of random noise. The resulting isotonic fits are shown as red lines. It can be seen that the isotonic regression line interpolates from a perfect fit to a random chance line. As long as there is not too much label noise, classification of the samples can be performed with reasonable accuracy by simply thresholding the regression line.

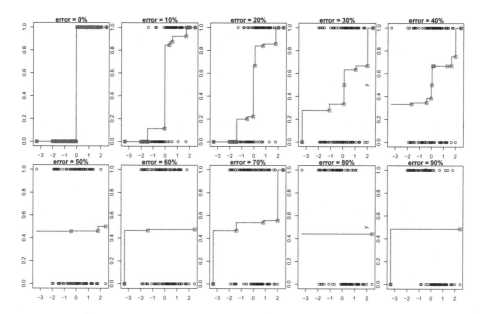

Fig. 3. An example of isotonic regression on an artificial data set with varying noise levels. This image was produced with the `isoreg` function from the R language [66].

The IVAP works by constructing two such regression lines for each test sample x with score $s \in \mathbb{R}$ assigned by the classifier: one where the calibration data is augmented with $(s, 0)$ and one where it is augmented with $(s, 1)$. The values of these curves at s form the lower and upper bounds respectively on the probability that the sample belongs to the positive class. For large data sets, we expect these values to be close together because the isotonic regressions are relatively insensitive to noise and outliers (as argued in [86]).

In prior work, we showed that it is possible to use IVAPs to detect adversarial manipulation of inputs for binary classification tasks [62]. Our goal here is to extend this work to the multiclass case and to construct an algorithm which provides stronger robustness and higher accuracy than the method we proposed previously.

3 MultIVAP

The IVAP is only designed for *binary* classification tasks and cannot directly work with multiclass problems. There exist two common methods for extending binary classifiers to the multiclass case, known as *one-vs-one (OvO)* and *one-vs-all (OvA)* [57]. To keep the computational overhead to a minimum, we opt for the OvA strategy here with a particular choice of aggregation method which will give us certain theoretical calibration guarantees. The full pseudocode is given in Algorithm 3, which we have called the *MultIVAP*.

Algorithm 3: MultIVAP

Input: bag of examples $\{z_1, \ldots, z_n\} \subseteq \mathcal{Z}$, object $x \in \mathcal{X}$, learning
algorithm A, significance level ε
Output: subset of labels from $\{1, \ldots, K\}$

1 Divide the bag of examples into a proper training set $\{z_1, \ldots, z_m\}$ and a
   calibration set $\{z_{m+1}, \ldots, z_n\}$.
2 Run the learning algorithm A on the proper training set to obtain a
   scoring classifier F.
3 **for** i *from 1 to* K **do**
4 | Let F_i be the score assigned by F to the ith class.
5 | Create a new calibration set $S_i = \{(x_{m+1}, y'_{m+1}), \ldots, (x_n, y'_n)\}$ where
   | $y'_j = \mathbb{1}[y_j = i]$.
6 | Use an IVAP to obtain probabilities $p_0^{(i)}, p_1^{(i)}$ for x using S_i and F_i for
   | calibration.
7 **end**
8 Solve (MultIVAP) to obtain an optimal solution $V \subseteq \mathcal{Y}$.
9 **return** V.

Consistent with the OvA strategy, the MultIVAP essentially works by fitting K IVAPs, one for each class, where the ith IVAP must decide whether a sample belongs to class i or not. The MultIVAP then takes the lower and upper bounds output by each of the IVAPs, $\left(p_0^{(1)}, p_1^{(1)}\right), \ldots, \left(p_0^{(K)}, p_1^{(K)}\right)$, and solves a mixed integer linear program (MILP) described below. The point of this optimization is to assign a (possibly empty) set of labels $V \subseteq \mathcal{Y}$ to the sample x such that certain probabilistic guarantees can be given on the result. Specifically, we consider a *multi-probabilistic* problem where instead of a single label Y there is a stochastic set L of possible labels that can be assigned to a given sample X. It is necessary to consider this generalization of the typical single-label classification setting, since it is known that perfect calibration *cannot* be achieved unless the classifier is multi-probabilistic [85]. That is, classification rules which output only a single label each time can never be perfectly calibrated. The probabilities computed by the IVAPs then bound the likelihood that a certain label should be included in this set:

$$p_0^{(i)} \leq \Pr[i \in L \mid X] \leq p_1^{(i)}. \tag{3}$$

Given a concrete realization x of the random variable X, we wish to construct a set of candidate labels $V \subseteq \mathcal{Y}$ which approximates the associated realization of L as well as possible. That is, we want to find the largest set of labels V such that the probability that each label $i \in V$ also belongs to L is maximized. Formally, this objective corresponds to maximizing $\Pr[V \subseteq L \mid X]$. To this end, we note that

$$\Pr[V \subseteq L \mid X] = \Pr\left[\bigwedge_{i \in V} (i \in L) \,\middle|\, X\right]$$

$$\overset{(a)}{\geq} \sum_{i \in V} \Pr[i \in L \mid X] - (|V| - 1)$$

$$\overset{(b)}{\geq} \sum_{i \in V} p_0^{(i)} - (|V| - 1).$$

The first inequality (a) follows from the intersection bound and the second inequality (b) follows from (3). Furthermore,

$$\Pr[V \subseteq L \mid X] \leq \min_{i \in V} \ \Pr[i \in L \mid X] \leq \min_{i \in V} \ p_1^{(i)}.$$

Combining the above results, we have the bounds

$$\sum_{i \in V} p_0^{(i)} - (|V| - 1) \leq \Pr[V \subseteq L \mid X] \leq \min_{i \in V} \ p_1^{(i)}. \tag{4}$$

Fix an $\varepsilon \in [0, 1]$ and let $\alpha_i = \mathbb{1}[i \in V]$. In order for $\Pr[V \subseteq L \mid X] \geq 1 - \varepsilon$ to hold, by (4) it is sufficient to have

$$\sum_{i \in V} p_0^{(i)} - (|V| - 1) \geq 1 - \varepsilon.$$

This is equivalent to

$$\sum_{i=1}^{K} \alpha_i \left(p_0^{(i)} - 1\right) \geq -\varepsilon. \tag{5}$$

In order to maximize the upper bound on $\Pr[V \subseteq L \mid X]$, it is necessary to maximize the smallest $p_1^{(i)}$ over all $i \in V$. That is,

$$\Pr[V \subseteq L \mid X] \leq \min_{i \in \mathcal{Y}} 1 + \alpha_i \left(p_1^{(i)} - 1\right). \tag{6}$$

Our objective is now formally to maximize the cardinality of V and the upper bound (6) while maintaining the constraint (5). This way, we obtain a conservative estimate of all labels that could be included in L with probability at least $1 - \varepsilon$. We can construct this estimate by solving the following MILP:

$$\begin{aligned}
& \underset{\alpha_1, \ldots, \alpha_K, q}{\text{maximize}} && \alpha_1 + \cdots + \alpha_K + q \\
& \text{subject to} && \alpha_1, \ldots, \alpha_K \in \{0, 1\}, \\
& && q \in [0, 1] && \text{(MultIVAP)}, \\
& && q + \alpha_i \left(1 - p_1^{(i)}\right) \leq 1, \quad i = 1, \ldots, K, \\
& && \sum_{i=1}^{K} \alpha_i \left(p_0^{(i)} - 1\right) \geq -\varepsilon.
\end{aligned}$$

Here, the binary variables α_i determine which labels get included in the solution V and q is a dummy variable which represents the smallest $p_1^{(i)}$ over all $i \in V$. Every solution V to (MultIVAP) satisfies the bounds

$$1 - \varepsilon \leq \Pr[V \subseteq L \mid X] \leq \begin{cases} \min_{i \in V} \ p_1^{(i)} & \text{if } V \text{ is non-empty,} \\ 1 & \text{otherwise.} \end{cases}$$

Note that at least one feasible solution to (MultIVAP) always exists, namely $\alpha_1 = \cdots = \alpha_K = 0$ and $q = 1$. This corresponds to taking the empty set $V = \emptyset$ as the prediction for L, which of course satisfies the trivial guarantee $\Pr[\emptyset \subseteq L \mid X] = 1$. If no solution exists other than the trivial empty set, then the prediction is rejected at the ε significance level. Figure 4 gives a high-level schematic overview of the inference phase of the MultIVAP.

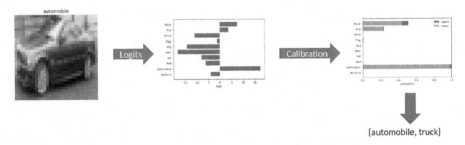

(a) A neural network takes the original sample (belonging to the class of automobiles) and outputs a logit vector. The MultIVAP makes use of ten IVAPs to convert the logits of these ten different classes into probability bounds. These bounds are then merged via an optimization method to yield a set of two labels: automobile and truck. This means that, based on the model output, our method believes that at least one of these two labels is appropriate for the given sample.

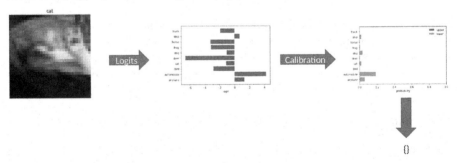

(b) In this example, the model is unreliable on the given input: it would predict automobile whereas the correct label is cat. Our method detects this and responds by not outputting any labels at all; the prediction is empty and the sample is rejected.

Fig. 4. Schematic overview of the inference phase of the MultIVAP.

3.1   Computational Complexity

There is still the question of the computational efficiency of Algorithm 3. Note that we can split the algorithm into two distinct stages:

1. *Calibration.* When the MultIVAP is first initialized, it is instantiated with a learning algorithm and a training data set. During calibration, it runs the learning algorithm and then runs the IVAP algorithm K times, once for each class. This step needs to be performed only once.
2. *Inference.* Once the MultIVAP is fully initialized, new samples can be processed using the precomputed isotonic regressions to obtain the probabilistic bounds. Then, the final prediction is computed by solving a MILP.

The complexity of the inference step is hard to quantify due to the solution of an optimization problem; we perform timing experiments in Sect. 4 in order to empirically estimate the overhead in this phase. It is easy to see, however, that the overhead incurred in the calibration step is $\mathcal{O}(Kc\log c)$ where c is the size of the calibration set. This follows because we fit K IVAPs and the complexity of the IVAP is dominated by the sorting step in the isotonic regressions, as discussed in [86].

3.2   Choosing the Calibration Set

Another important aspect of the MultIVAP is the question of how to choose the calibration set. Formally, the only requirement is that this set must be an independent sample from the data distribution similar to the test or validation set [86]. However, it would be interesting to have finite-sample guarantees on the performance of the predictor depending on the size c of the calibration set. To the best of our knowledge, such guarantees do not exist yet. [86] contains the only real discussion we have been able to find in the literature on this issue. There, the authors state that the lower and upper bounds $(p_0^{(i)}, p_1^{(i)})$ used by the MultIVAP will lie very close together for large data sets, but this is only an asymptotic statement that says little about the actual rate of convergence when the data set is finite.

In our own work [33], we have found that the IVAP can be sensitive to the particular choice of calibration set size to the point where the predictions can sometimes become unusable. Unsurprisingly, this appears to depend on the specifics of the underlying model; that is, some models are harder for the IVAP to calibrate than others. In the absence of theoretical finite-sample guarantees, the size of the calibration set may be treated as another hyperparameter which can be tuned on a separate validation set. The optimal size can then be chosen depending on how well the accuracy of the resulting predictor is balanced against other considerations such as computational overhead.

3.3 Defense-Aware White-Box Attack

To enable a fair and thorough evaluation of the MultIVAP, we also design a custom defense-aware white-box attack specifically for fooling this defense[5]. Given that the MultIVAP is a multi-probabilistic predictor returning a set of labels rather than a single prediction, we must slightly modify the typical goal of causing a misclassification since that is ill-defined here. On the one hand, we do not want the correct label to be present in the resulting label set; on the other hand, we do not want the MultIVAP to output too many or too few labels either as this can also raise suspicion. We settle on the following optimization problem:[6]

$$\min_{\tilde{x} \in \mathcal{X}} \|x - \tilde{x}\|_\infty + \lambda \|F(\tilde{x}) - s\|_\infty. \tag{7}$$

Here, $F : \mathcal{X} \to \mathbb{R}^K$ is the scoring classifier, $\lambda \in \mathbb{R}$ is a parameter and $s \in \mathbb{R}^K$ is a target vector of scores. The scalar λ is optimized via binary search so that the magnitude of the perturbation $\|x - \tilde{x}\|_\infty$ is as small as possible. The score vector s is determined by searching the calibration set of the MultIVAP for all calibration samples (x', y') such that y' does not belong to the prediction region of x. Among these candidates, we randomly sample one element x' and take its calibration score vector as the target $s = F(x')$.

An adversarial example produced by (7) is accepted only if the following conditions are met:

1. The perturbation stays within the budget, that is, $\|x - \tilde{x}\|_\infty \le \eta$ where η is a user-specified perturbation bound.
2. The MultIVAP does not return an empty prediction region for the adversarial sample.
3. The true label is not present in the prediction region returned by the MultIVAP.

Solutions to (7) satisfying these properties are counted as "successes", and the *success rate* of the attack is the fraction of samples that were counted as such.

The intuition behind our defense-aware attack is the following. The K IVAPs used internally by the MultIVAP will output the same upper and lower bounds for all samples that are assigned the same scores by the classifier F. Therefore, if we can corrupt a sample x into a sample $\tilde{x}$ which shares the score vector of another sample x', then $\tilde{x}$ will be associated with the same probabilities as x' and the optimization problem (MultIVAP) will have the same solution sets for both. This will cause the MultIVAP to output the same prediction regions for $\tilde{x}$ and x'. If, furthermore, the distance between x and $\tilde{x}$ is small and if the true label of x is not present in the prediction region of x', then $\tilde{x}$ can be considered an adversarial example. This is what we aim to achieve in (7). In finding an

---

[5] See [11] for an overview of the various desiderata that an adversarial defense evaluation should satisfy.

[6] We use the ℓ_∞ norm everywhere as this is recommended by [51]. However, the attack can be trivially adapted to any other norm.

appropriate sample x' to target, we try to minimize $\|F(x) - F(x')\|_\infty$ in order to "warm-start" the attack.

Note additionally that the objective (7) can be optimized using gradient descent without issue, as the scoring classifier F is a standard (fully differentiable) neural network. In particular, our defense-aware attack will not suffer from any so-called *gradient masking* [4]. This is a common problem affecting gradient-based adversarial attacks where a proposed defense method has "masked" or somehow corrupted the gradient signal of the model it is supposed to defend. As shown by [4], this does not actually make models more robust. It can, however, create the illusion of robustness if one only evaluates against gradient-based attacks. Importantly, standard neural networks trained using typical methods generally do not exhibit gradient masking, making it not a concern for our attack.

4   Experiments

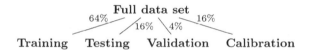

Fig. 5. Illustration of the different data splits used in our experiments.

We perform experiments on MNIST, Fashion-MNIST, CIFAR-10, Asirra and SVHN data sets [23,47,48,60,88]. For each data set, we normalize the pixel values to the interval $[0, 1]$. For the Asirra data set, we also resized all images to 64×64 pixels to facilitate processing by our pipeline as the images come in various irregular sizes. The data splits we used are illustrated in Fig. 5: we used 64% of the data for training the neural networks, 16% for testing, 4% for validation and 16% for calibrating the MultIVAP. These splits were constructed by randomly shuffling the entire data set and subdividing the samples accordingly.

We train convolutional neural networks and apply Algorithm 3 to obtain a calibrated multi-probabilistic predictor. The exact architectures used for each data set are described in appendix A, listings A.1 to A.5. We used the Keras library with the TensorFlow backend for training the neural networks [1,15]. Each network was optimized using the Adam optimizer [45] and trained for 10 epochs, except for Asirra which was trained for 50 epochs. For the IVAPs, we made use of an implementation by [82]. We report several metrics to assess the performance of the MultIVAP:

- The accuracy (ACC) of the original model. We give both the accuracy of the original model on the full test set as well as the accuracy of the model on the subset of predictions accepted by the MultIVAP. This is referred to as the "corrected accuracy" (COR).

- The average *Jaccard index* (JAC) of the MultIVAP on the test set [39]. This index, also known as the *intersection over union* (IoU), is defined as

$$J(x) = \frac{|L(x) \cap V(x)|}{|L(x) \cup V(x)|}$$

where $L(x)$ is the true label set for the sample x and $V(x)$ is the MultIVAP prediction region. In the special case where $L(x)$ is a singleton (single-label prediction), this formula reduces to

$$J(x) = \begin{cases} 1/|V(x)| & \text{if } L(x) \subseteq V(x), \\ 0 & \text{otherwise.} \end{cases}$$

That is, it is inversely proportional to the size of the prediction region $V(x)$ if this region contains the correct label. If it does not, then a score of zero is assigned. In particular, rejected predictions where $V(x) = \emptyset$ are treated as if they are erroneous. The average Jaccard index over the test set is then given by

$$\overline{J} = \frac{1}{m} \sum_{i:L(x_i) \subseteq V(x_i)} \frac{1}{|V|}.$$

If the MultIVAP always returns exactly one label, then the average Jaccard index coincides with the typical notion of test accuracy. In general, however, the MultIVAP can return multiple labels or even no labels at all in case of rejection. Therefore, $\overline{J}$ will always be bounded by the accuracy and it is expected that it will often be smaller.
- The *predictive efficiency* (EFF), which is the average number of labels the MultIVAP outputs across all samples. Ideally, this number should be very close to one.
- The significance level ε at which the results for the MultIVAP were obtained. This level was determined by tuning the threshold on a held-out validation set such that the predictive efficiency was as close to one as possible.
- The true rejection rate (TRR) along with the false rejection rate (FRR) and the overall rejection rate (REJ). Formally, these are computed as follows:

$$\text{TRR} = \frac{\text{TR}}{\text{TR} + \text{FA}}, \quad \text{FRR} = \frac{\text{FR}}{\text{FR} + \text{TA}}, \quad \text{REJ} = \frac{\text{TR} + \text{FR}}{\text{TR} + \text{FR} + \text{TA} + \text{FA}}.$$

The quantities appearing in these equations are the number of true rejections (TR), false acceptances (FA), false rejections (FR) and true acceptances (TA). Samples are rejected if the MultIVAP returns an empty prediction region. A rejection is true if the underlying scoring classifier F would have returned an erroneous prediction and false otherwise.

It is of particular importance for us to study the Jaccard index, efficiency, TRR and FRR of the MultIVAP together and not simply the accuracy score. This is because the MultIVAP is a *multi-label* predictor, meaning it yields a set of possible labels for each sample instead of a single point prediction like most models do. When considering the performance of such a method compared to a standard classifier that outputs an argmax over a set of estimated probabilities, we must ask ourselves the following questions:

1. How well do the label sets predicted by the MultIVAP correspond to the true label sets? This is measured by the Jaccard index, which is one possible generalization of the accuracy score specifically for multi-label predictors.
2. How many labels does the MultIVAP output on average? This is the predictive efficiency. We do not want this quantity to be much higher or lower than the ground truth (which is usually one), otherwise the MultIVAP can create confusion.
3. When the MultIVAP rejects a prediction, what is the probability that this rejection was "justified", in the sense that the underlying model used by the MultIVAP was wrong? This is quantified by the true rejection rate and the false rejection rate.

Naturally, we want the Jaccard index at 100%, the predictive efficiency close to one, the TRR close to 100% and the FRR close to 0%. We can exercise some amount of control over these metrics by varying the significance level ε. In our experiments, we used a held-out validation set to tune ε in order to have a predictive efficiency as close to one as possible. Depending on the particular application, however, one can use many other methods to tune ε. For instance, in cybersecurity settings, it is considered highly undesirable to unjustly flag benign samples as malicious because this interferes with legitimate users' business. In such cases, we might be more concerned with minimizing the FRR than we are with maximizing predictive efficiency (or we may want to specify some sort of trade-off between these two metrics) and so we might tune ε so that the FRR does not exceed some specified threshold instead.

To assess the robustness of the MultIVAP against adversarial perturbations, we consider two threat models:

Defense-Oblivious. Here, we evaluate the defense against non-adaptive transfer attacks. That is, the attacks have no knowledge of the defense; they are crafted against the baseline model and then simply presented to the MultIVAP. This is the bare minimum of robustness that any defense should hope to achieve [11]. The attacks we tested against here are DeepFool [56], projected gradient descent (PGD; [51]), fast gradient sign method (FGSM; [31]), single pixel [77] and local search [59]. All implementations were provided by the Foolbox library [68]. We chose this particular set of attacks because of their diversity: there are gradient-based attacks (DeepFool, PGD and FGSM), non-gradient-based attacks (Single pixel and local search), iterative (DeepFool, PGD, single pixel, local search) and single step attacks (FGSM).

Defense-Aware. In this setting, we use our own adaptive attack detailed in Sect. 3 to specifically bypass the MultIVAP. This evaluation should be a worst-case scenario for our defense and provides an estimation of a lower bound on its actual robustness.

Finally, we address the question of computational efficiency of our method. It is difficult to precisely quantify the complexity of the MultIVAP as it requires the solution of a MILP, which is NP-complete in general [7]. As such, we perform timing experiments where we measure the average time per prediction for both the baseline model and the MultIVAP on the test sets of each task.

The code for all our experiments is available at https://github.com/saeyslab/multivap. We consistently used 64% of the data for training, 16% for testing, 4% for validation and 16% for calibration.

4.1   Results

Table 1. Results of the baseline models and the MultIVAPs on the different data sets.

Task	Baseline	MultIVAP					
	ACC (COR)	JAC	$1 - \varepsilon$	EFF	TRR	FRR	REJ
MNIST	99.10% (99.59%)	94.16%	33.75%	0.99	56.94%	4.54%	2.90%
FMNIST	91.81% (93.84%)	83.44%	24.20%	1.04	29.62%	4.45%	6.75%
CIFAR-10	76.40% (81.52%)	61.86%	20.77%	1.05	32.73%	8.36%	15.54%
Asirra	88.82% (89.04%)	83.65%	41.86%	1.00	2.69%	0.48%	1.10%
SVHN	92.22% (96.40%)	80.77%	25.23%	1.01	59.22%	7.70%	10.42%

Results on clean data are reported in Table 1. We can see from the corrected accuracy that the MultIVAP can increase the accuracy of the original model at the cost of rejecting a certain percentage of the predictions. These percentages can vary significantly depending on the underlying model and the data set: on Asirra, we reject 1.10% of predictions whereas on CIFAR-10 we reject 15.54%. If one counts rejection as misclassification, then the corrected accuracy minus rejections (COR - REJ) can sometimes be lower than the baseline accuracy (ACC). However, depending on the specific application, it may not be appropriate to treat rejections as if they are errors: if correctness of the output is of paramount importance—such as when machine learning is used for medical applications, industrial control systems or autonomous vehicles—then it can be preferable to reject a relatively large number of predictions and return control to human moderators instead of allowing the model to continue functioning based on erroneous output. The risk of false rejection can be outweighed by the risks associated with misclassification, especially if the false rejection rate is relatively low (as it is in Table 1).

The Jaccard indices are relatively high, although (as expected) it is lower than the baseline accuracy because of the rejected predictions. The predictive efficiency is very close to one on all tasks, with high TRRs and relatively low FRRs. The significance levels $1 - \varepsilon$ are rather low, however, meaning the statistical guarantees provided by our method are relatively weak. This may be due to our specific way of tuning ε or it may be due to looseness of the bounds. Improving these bounds and obtaining better guarantees is left to future work.

Since the MultIVAP inevitably returns multiple labels sometimes, it is interesting to study which labels it tends to output together. This gives us an idea of the classes that the MultIVAP has difficulty separating and may be a useful tool for diagnosing model shortcomings. Figure 6 plots co-occurence matrices as heatmaps for each classification task, where the cell at row i, column j contains the number of times that labels i and j were returned together in a prediction on the proper test set. The diagonals are left blank for clarity since every label trivially co-occurs with itself each time. We see from these heatmaps that the classes that are most difficult for the MultIVAP to separate are also the most

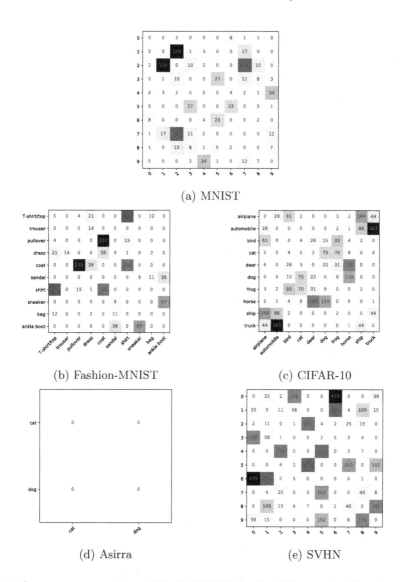

Fig. 6. Co-occurence matrices of the MultIVAP for the different classification tasks.

perceptually similar ones: on CIFAR-10, automobiles are confused with trucks and deers are confused with horses, for example. On the other hand, birds are never confused with horses and frogs are never confused with trucks. Similar results hold for the other data sets.

Figure 7 plots the accuracy (defined as the fraction of prediction regions containing the correct label), predictive efficiency and rejection rate of the MultI-VAP as a function of the significance level ε. The tuned value for ε is shown as a

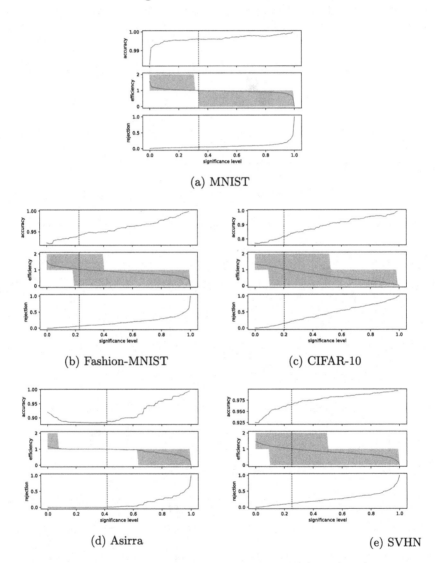

(a) MNIST

(b) Fashion-MNIST

(c) CIFAR-10

(d) Asirra

(e) SVHN

Fig. 7. Accuracy, predictive efficiency and rejection rate for the MultIVAP on the different tasks. Note that we shade the area between the 5th and 95th percentiles of the predictive efficiency.

dotted black line. As expected, a higher significance level yields higher accuracy and better predictive efficiency but also a higher rejection rate.

Figure 8 presents ROC curves of the MultIVAP for each task, where the false rejection rate is plotted against the true rejection rate. The tuned threshold ε is indicated with a dotted black line. We see that the MultIVAP achieves consistently high AUC scores, indicating that it is indeed capable of determining whether a prediction from the underlying model is reliable or not.

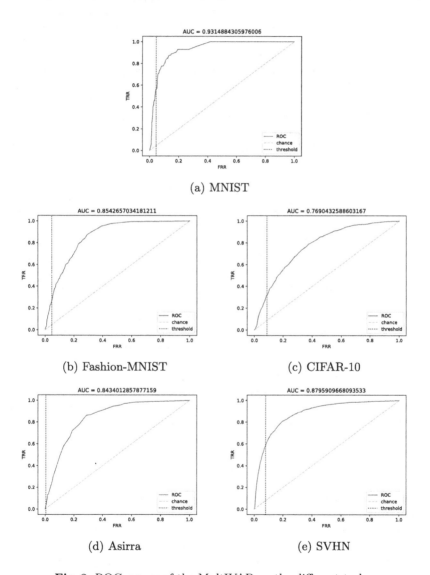

Fig. 8. ROC curves of the MultIVAP on the different tasks.

Results for adversarial transfer attacks are presented in Table 2. For MNIST and Fashion-MNIST, we chose $\eta = 0.3$; for all other data sets, we let $\eta = 0.03$ as these bounds are fairly typical in the literature. We can see that the MultIVAP is significantly more accurate than the original models and that it generally has a high TRR on adversarial samples.

Table 2. Performance metrics of the baseline models and the MultIVAPs on adversarial transfer attacks.

Task	Attack	Baseline	MultIVAP				
		ACC (COR)	JAC	EFF	TRR	FRR	REJ
MNIST ($\eta = 0.3$)	DeepFool	35.86% (35.83%)	27.18%	0.52	59.23%	60.86%	59.81%
	PGD	0.38% (0.33%)	37.23%	0.64	52.32%	46.67%	52.30%
	FGSM	4.39% (4.57%)	41.13%	1.10	39.64%	2.85%	38.03%
	Single pixel	98.01% (99.01%)	93.96%	0.99	62.89%	4.04%	5.21%
	LocalSearch	25.48% (26.13%)	43.24%	0.41	84.02%	2.60%	63.28%
FMNIST ($\eta = 0.3$)	DeepFool	27.98% (28.14%)	29.32%	0.66	43.60%	39.90%	42.56%
	PGD	0.00% (0.00%)	35.19%	0.69	41.59%	—	41.59%
	FGSM	1.10% (1.11%)	28.27%	0.53	56.76%	2.38%	56.19%
	Single pixel	85.05% (88.25%)	83.66%	1.02	35.28%	3.45%	8.21%
	LocalSearch	51.10% (52.46%)	73.33%	0.76	60.46%	2.84%	31.01%
CIFAR-10 ($\eta = 0.03$)	DeepFool	0.01% (0.01%)	37.47%	0.95	24.16%	16.10%	24.10%
	PGD	0.00% (0.00%)	35.28%	0.91	28.13%	—	28.13%
	FGSM	0.00% (0.00%)	32.38%	0.83	32.36%	0.00%	32.25%
	Single pixel	60.68% (68.12%)	59.43%	1.01	37.53%	4.39%	17.43%
	LocalSearch	18.99% (21.39%)	52.00%	0.73	48.62%	4.10%	40.16%
Asirra ($\eta = 0.03$)	DeepFool	0.00% (0.00%)	37.01%	0.99	1.18%	12.50%	12.04%
	PGD	0.00% (0.00%)	41.37%	0.98	1.60%	—	1.60%
	FGSM	0.00% (0.00%)	37.71%	0.98	1.64%	0.00%	1.63%
	Single pixel	75.40% (75.50%)	77.01%	1.00	1.63%	0.00%	0.40%
	LocalSearch	57.80% (57.87%)	73.14%	1.00	1.49%	0.00%	0.63%
SVHN ($\eta = 0.03$)	DeepFool	0.01% (0.01%)	31.69%	0.82	36.00%	16.34%	35.86%
	PGD	0.00% (0.00%)	27.98%	0.80	41.33%	0.00%	41.33%
	FGSM	0.00% (0.00%)	19.16%	0.54	58.72%	0.00%	58.55%
	Single pixel	69.59% (77.38%)	72.00%	0.96	58.50%	1.99%	15.78%
	LocalSearch	22.91% (25.47%)	56.55%	0.79	71.61%	2.01%	30.32%

Figure 9 shows the success rate of our defense-aware adversarial attack as a function of the number of iterations of gradient descent. We used the same values of η for the defense-aware attack as we did for the transfer attacks. The maximum number of iterations was limited to 5,000 as we noticed that the success rates for all tasks seemed to plateau after this point. The highest rate we were able to achieve on any task is less than 40%. For CIFAR-10 in particular, we have a success rate of less than 30%. By contrast, at the time of this writing, state of the art robust accuracy on CIFAR-10 is less than 50% for $\eta = 0.03$ [71]. On MNIST and Fashion-MNIST, our success rates of 15% and 30% at $\eta = 0.3$ seem to be competitive with existing methods, which achieve robust accuracies of 87.54% and 76.83% respectively [2]. Note also that our ℓ_∞ bound of 0.3 for Fashion-MNIST is three times higher than the 0.1 bound used by [2].

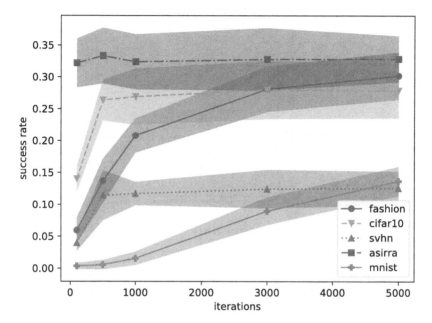

Fig. 9. Success rate of the defense-aware adversarial attack as a function of the number of iterations of optimization. The lines represent the mean success rate based on ten random restarts of the attack. The area between the 5th and 95th percentiles is shaded.

To gain more insights into the precise nature of our defense-aware adversarial examples, we plot the associated confusion matrices in Fig. 10. We see that these bear a striking similarity to the co-occurrence matrices from Fig. 6, where we observed that the MultIVAP mainly confuses perceptually similar classes. This means that when the defense-aware adversarial attack succeeds in fooling the MultIVAP, the resulting adversarial sample is usually misclassified into a class that is visually close to the original correct class. This is an important finding for two reasons:

– It represents a definite improvement over the state of the art in adversarial defense. Currently, even the most robust models can still be tricked into misclassifying a given sample into almost any other class regardless of similarity. The fact that the MultIVAP mostly confuses only perceptually similar classes is a good indicator that it is fundamentally more robust than other models.
– The mistakes that the MultIVAP makes on the adversarial examples are similar to the ones it makes on the test set. This suggests that we can make the model more robust via "traditional" means, that is, by decreasing test error using data augmentation or other tricks. Co-occurence matrices such as

Fig. 6 can show us which classes are the most difficult for the MultIVAP to separate. This information can be used to guide efforts for reducing test error. More advanced and computationally expensive methods such as adversarial training may not be required to increase robustness; standard techniques for improving the generalization performance might already suffice.

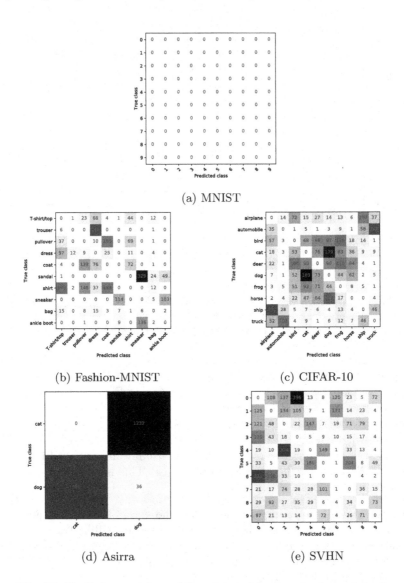

Fig. 10. Confusion matrices of the adversarial examples for each task.

Figure 11 shows the results of our timing experiments. Here, we compute the average time per prediction for the baselines as well as the MultIVAPs and calculate the base-10 log of the ratio between these two quantities over the proper test set:

$$\mathrm{LogRatio} = \log_{10}\left(\frac{\text{average MultIVAP inference time}}{\text{average baseline inference time}}\right).$$

We find that the log-ratio is close to one for problems with ten classes whereas it is close to 0.2 for the binary classification task. This translates to a slow-down factor of ten and 1.6 respectively. In practice, we found that this overhead was negligible in wall clock time for the models and data sets we tested.

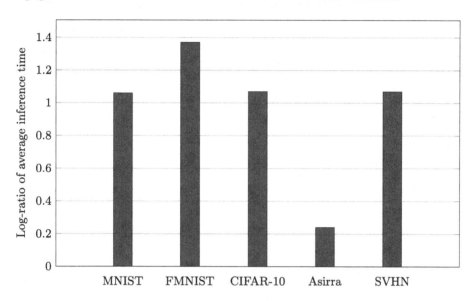

Fig. 11. Comparison of the computational efficiency of the baseline models and the MultIVAPs. These measurements were taken on a single machine utilizing one NVIDIA Titan Xp GPU alongside four Intel Core i5-6600 CPUs @ 3.30 GHz each.

4.2   Comparison with Adversarial Training

At present, adversarial training seems to be the most successful defense against adversarial perturbation known in the literature. In particular, the ℓ_∞ PGD adversarial training method by [51] is considered state of the art in this field. However, it introduces a considerable computational overhead: it can easily increase the time required to fully train a given model by a factor of 30. Recently, [71] proposed a novel "free" adversarial training procedure which appears to achieve similar results to the method of [51] while keeping the computational overhead negligible compared to standard training. Therefore, we have opted to compare our MultIVAP algorithm to the adversarial training method by [71]. The pseudo-code is shown in Algorithm 4. This algorithm is identical to standard minibatch stochastic gradient descent training for neural networks, except that a single "running perturbation" δ is stored and constantly updated at the end of each iteration. This perturbation is used to create the adversarial minibatches. In standard PGD training, these minibatches are created by re-running the PGD optimization from scratch every iteration. Free adversarial training speeds this up by replacing this expensive optimization by a single gradient update to a single perturbation δ. Multiple gradient updates are supported (and in fact recommended) via the hop step parameter m, which must be set to a divisor of the number of epochs in order to keep the number of gradient updates to the model parameters constant.

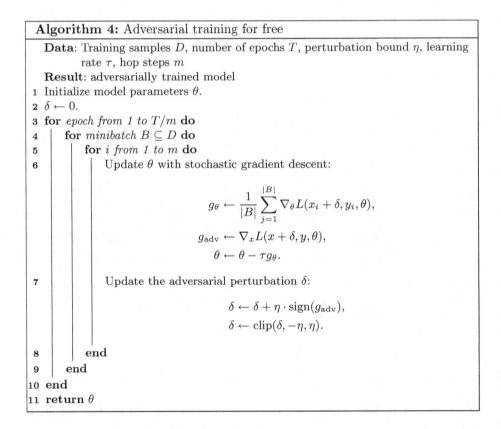

Algorithm 4: Adversarial training for free

Data: Training samples D, number of epochs T, perturbation bound η, learning rate τ, hop steps m

Result: adversarially trained model

1 Initialize model parameters θ.
2 $\delta \leftarrow 0$.
3 **for** *epoch from 1 to T/m* **do**
4   **for** *minibatch $B \subseteq D$* **do**
5   **for** *i from 1 to m* **do**
6   Update θ with stochastic gradient descent:

$$g_\theta \leftarrow \frac{1}{|B|} \sum_{j=1}^{|B|} \nabla_\theta L(x_i + \delta, y_i, \theta),$$
$$g_{\text{adv}} \leftarrow \nabla_x L(x + \delta, y, \theta),$$
$$\theta \leftarrow \theta - \tau g_\theta.$$

7   Update the adversarial perturbation δ:

$$\delta \leftarrow \delta + \eta \cdot \text{sign}(g_{\text{adv}}),$$
$$\delta \leftarrow \text{clip}(\delta, -\eta, \eta).$$

8   **end**
9   **end**
10 **end**
11 **return** θ

To make our comparison, we retrain our baseline models used by the MultIVAP with the free adversarial training method[7] and subject these models to the ℓ_∞ PGD adversarial attack. The data preprocessing, perturbation bounds as well as the number of training epochs used here were all kept identical to the

---

[7] Implementation available at https://github.com/ashafahi/free_adv_train. Accessed 2020-06-17.

ones in the previous sections. We set the number of hop steps to ten in all our experiments, since this number achieved good results according to [71] and is a divisor of our total number of epochs (10 or 50 depending on the data set).

Before we present our experimental comparison with free adversarial training, however, it is important to note that our defense is relatively unique in the field of adversarial ML since it is multi-valued. Most approaches to adversarial robustness focus on *hardening* classifiers so that their misclassification rate when exposed to adversarial samples is reduced. The resulting models typically still return point predictions and as such they are evaluated using the standard accuracy metric. The MultIVAP, being multi-valued, is evaluated instead using a combination of different metrics, including the Jaccard index (a generalization of standard accuracy) and the rejection rates. We have done our best to provide a fair comparison between the MultIVAP and free adversarial training, but we believe the choice is ultimately determined by the preferences of the practitioner and cannot be reduced to any single objective metric.

Table 3. Accuracy scores of the adversarially trained models on clean data as well as adversarial examples generated by the ℓ_∞ PGD attack.

Task	Clean	Robust
MNIST	97.36%	93.03%
FMNIST	83.27%	74.64%
CIFAR-10	23.54%	15.94%
Asirra	50.00%	50.00%
SVHN	19.59%	19.59%

The results of our experiments with free adversarial training are shown in Table 3. It is immediately clear that these numbers are far inferior to those obtained by [71]. This can be attributed to the choice of model: we used very simple CNNs (see Appendix A) whereas Shafahi et al. made use of ResNet architectures [36]. On some data sets such as CIFAR-10, SVHN and Asirra, free adversarial training using our baseline models does not perform significantly better than random guessing and the MultIVAP is clearly superior. On MNIST and Fashion-MNIST, however, we do note a significant increase in adversarial robustness against the ℓ_∞ PGD attack. On MNIST, the clean and robust accuracies of free adversarial training are higher than the corresponding Jaccard indices of the MultIVAP. For Fashion-MNIST, the clean accuracy is comparable to the Jaccard index of the MultIVAP but the robust accuracy is higher. Note, however, that for both of these data sets the MultIVAP rejects a large portion of adversarial examples which would have fooled the model: the true rejection rate exceeds 50% for MNIST and 40% for Fashion-MNIST.

We conclude from this that our method has an interesting comparative advantage with respect to free adversarial training: the MultIVAP can already provide significant adversarial robustness for simple models whereas free adversarial training requires more complex models in order to outperform the MultIVAP on the same data set. This is in line with existing research on adversarial robustness which indicates that in order to obtain more robust models one must increase model capacity [58]. With the MultIVAP approach, this increase in model complexity is avoided since the burden of detecting adversarial examples is carried by the conformal predictor instead. Of course, there is a price to pay for this advantage: free adversarial training by itself incurs virtually no overhead at all—although, arguably, this overhead is merely "hidden" in the additional model complexity required to achieve significant robustness—whereas the MultIVAP incurs a cost in both the training phase as well as the testing phase. As has already been discussed, the overhead of training the MultIVAP is log-linear in the size of the calibration set. The overhead in the testing phase depends on the number of classes and is visualized in Fig. 11. A precise characterization of the computational complexity of this phase is difficult since it involves the solution of a MILP, but in our experiments the overhead was negligible since the baseline models were already very efficient to evaluate.

These results suggest an important topic for future work: a large-scale benchmark study which characterizes the trade-off between model complexity and adversarial robustness for different defense methods. The main question of interest to us is whether the MultIVAP is more "model efficient" than existing state of the art adversarial defenses, in the sense that it can match or outperform the robustness provided by other methods using simpler models. The experiments we performed here indicate that the MultIVAP is indeed more efficient on the models and data sets we tested, but a complete picture would require us to compare the MultIVAP to adversarial training on larger data sets such as ImageNet [20] and bigger models such as Wide ResNet architectures [90].

5   Conclusion

We have proposed a computationally efficient multiclass generalization of the inductive Venn-ABERS prediction algorithm which we have called the *MultIVAP*. In our experiments, we have found that this method significantly increases both the accuracy as well as the adversarial robustness of any model with which it is instantiated. The MultIVAP takes two hyperparameters, a calibration data set and a significance level ε, which can be tuned according to various objectives. The method enjoys the theoretical guarantee that its label sets will contain the true label with a probability of at least $1 - \varepsilon$. The code for our implementation is available at https://github.com/saeyslab/multivap.

Several avenues for future work are possible. For one, the probabilistic bounds we have obtained for the error probability of our method have the advantage that they reduce inference using the MultIVAP to a MILP, which can be solved relatively efficiently. However, this efficiency comes at the cost of looseness: the tuned values of $1 - \varepsilon$ we computed in our experiments are very low (always less than 50% and often less than 25%). Obtaining tighter probabilistic bounds with stronger guarantees on the label set is perhaps the most interesting improvement that could be made. However, the resulting optimization problem might no longer be linear or even convex, making the problem more difficult to solve.

Improving the computational efficiency of this method when the number of classes becomes very high is definitely a worthwhile avenue for future work as well. Currently, the MultIVAP appears to increase inference time by a factor that is approximately equal to the number of classes. In applications that are very time-sensitive and which have a large number of classes (e.g. extreme classification [40]), this might be unacceptable. The main bottleneck here is the optimization problem; the other part of the algorithm which gathers the outputs of the different IVAPs is straightforward to parallelize as the different classes can be treated independently.

Furthermore, in data-limited settings, it may be undesirable to sacrifice a relatively large portion of the data set for calibration of the MultIVAP, especially if a separate validation split is required for tuning other hyperparameters. We did not study the effect of the size of the calibration set in detail here, but it can prove useful to test how small this set can be made before the MultIVAP is unable to meet certain guarantees on robustness and accuracy.

It may also be possible to develop stronger defense-aware attacks against the MultIVAP, since we only studied one attack in this work. We recall *Schneier's Law* [70]:

> Anyone, from the most clueless amateur to the best cryptographer, can create an algorithm that they themselves cannot break.

We therefore invite the community to scrutinize our defense and develop stronger attacks against it. Nevertheless, even if our particular defense is ever broken, we believe there is a broader point to be made here: to the best of our knowledge, there exist almost no papers that attempt to tackle adversarial machine learning using techniques from the field of conformal prediction. Given that this entire field is dedicated precisely to the problem of trustworthiness of probabilistic predictions, it would seem there is still much unexplored potential in this area. We refer the interested reader to the works of [72,85,86].

A   Network Architectures

```
1   input_shape = (28, 28, 1)
2   model = Sequential()
3   model.add(Conv2D(32, kernel_size=(5, 5),
4                    activation='relu',
5                    input_shape=input_shape))
6   model.add(Conv2D(64, (5, 5), activation='relu'))
7   model.add(MaxPooling2D(pool_size=(2, 2)))
8   model.add(Flatten())
9   model.add(Dense(128, activation='relu'))
10  model.add(Dense(num_classes))
11  model.add(Activation('softmax'))
```

Listing A.1: Network architecture for MNIST

```
1   input_shape = (28, 28, 1)
2   model = Sequential()
3   model.add(Conv2D(32, kernel_size=(3, 3),
4                    activation='relu',
5                    input_shape=input_shape))
6   model.add(Conv2D(32, (3, 3), activation='relu'))
7   model.add(MaxPooling2D(pool_size=(2, 2)))
8   model.add(Dropout(.25))
9   model.add(Conv2D(64, (3, 3), activation='relu'))
10  model.add(Conv2D(64, (3, 3), activation='relu'))
11  model.add(MaxPooling2D(pool_size=(2, 2)))
12  model.add(Dropout(.25))
13  model.add(Flatten())
14  model.add(Dense(512, activation='relu'))
15  model.add(Dropout(.5))
16  model.add(Dense(num_classes))
17  model.add(Activation('softmax'))
```

Listing A.2: Network architecture for Fashion-MNIST

```
1   input_shape = (32, 32, 3)
2   inp = Input(shape=input_shape)
3   x = Conv2D(32, (3, 3), padding='same', input_shape=input_shape)(inp)
4   x = Activation('relu')(x)
5   x = Conv2D(32, (3, 3))(x)
6   x = Activation('relu')(x)
7   x = MaxPooling2D(pool_size=(2, 2))(x)
8   x = Dropout(.2)(x)
9
10  x = Conv2D(64, (3, 3), padding='same')(x)
11  x = Activation('relu')(x)
12  x = Conv2D(64, (3, 3))(x)
13  x = Activation('relu')(x)
14  x = MaxPooling2D(pool_size=(2, 2))(x)
15  x = Dropout(.2)(x)
16
17  x = Flatten()(x)
18  x = Dense(512)(x)
19  x = Activation('relu')(x)
20  x = Dropout(.2)(x)
21  x = Dense(num_classes)(x)
22  x = Activation('softmax')(x)
```

Listing A.3: Network architecture for CIFAR-10

```
1   input_shape = (32, 32, 3)
2   model = Sequential()
3   model.add(Conv2D(32, kernel_size=(3, 3),
4                    activation='relu',
5                    input_shape=input_shape))
6   model.add(Conv2D(32, (3, 3), activation='relu'))
7   model.add(MaxPooling2D(pool_size=(2, 2)))
8   model.add(Dropout(.25))
9   model.add(Conv2D(64, (3, 3), activation='relu'))
10  model.add(Conv2D(64, (3, 3), activation='relu'))
11  model.add(MaxPooling2D(pool_size=(2, 2)))
12  model.add(Dropout(.25))
13  model.add(Flatten())
14  model.add(Dense(512, activation='relu'))
15  model.add(Dropout(.5))
16  model.add(Dense(num_classes))
17  model.add(Activation('softmax'))
```

Listing A.4: Network architecture for SVHN

```
1    input_shape = (64, 64, 3)
2    num_classes = 2
3    model = Sequential()
4    model.add(Conv2D(32, kernel_size=(3, 3),
5                     activation='relu',
6                     input_shape=input_shape))
7    model.add(Conv2D(32, (3, 3), activation='relu'))
8    model.add(MaxPooling2D(pool_size=(2, 2)))
9    model.add(Dropout(.25))
10   model.add(Conv2D(64, (3, 3), activation='relu'))
11   model.add(Conv2D(64, (3, 3), activation='relu'))
12   model.add(MaxPooling2D(pool_size=(2, 2)))
13   model.add(Dropout(.25))
14   model.add(Conv2D(128, (3, 3), activation='relu'))
15   model.add(Conv2D(128, (3, 3), activation='relu'))
16   model.add(MaxPooling2D(pool_size=(2, 2)))
17   model.add(Dropout(.25))
18   model.add(Flatten())
19   model.add(Dense(512, activation='relu'))
20   model.add(Dropout(.5))
21   model.add(Dense(num_classes))
22   model.add(Activation('softmax'))
```

Listing A.5: Network architecture for Asirra

References

1. Abadi, M., et al.: TensorFlow: a system for large-scale machine learning. In: 12th USENIX Symposium on Operating Systems Design and Implementation (OSDI 16), pp. 265–283 (2016)
2. Andriushchenko, M., Hein, M.: Provably robust boosted decision stumps and trees against adversarial attacks. In: Advances in Neural Information Processing Systems, pp. 12997–13008 (2019)
3. Arora, S., Ge, R., Neyshabur, B., Zhang, Y.: Stronger generalization bounds for deep nets via a compression approach. In: Dy, J., Krause, A. (eds.) Proceedings of the 35th International Conference on Machine Learning. Proceedings of Machine Learning Research, Stockholmsmässan, Stockholm, Sweden, vol. 80, pp. 254–263. PMLR, 10–15 July 2018. http://proceedings.mlr.press/v80/arora18b.html
4. Athalye, A., Carlini, N., Wagner, D.: Obfuscated gradients give a false sense of security: circumventing defenses to adversarial examples. arXiv preprint arXiv:1802.00420 (2018)
5. Azuma, K.: Weighted sums of certain dependent random variables. Tohoku Math. J. Second Ser. **19**(3), 357–367 (1967)
6. Biggio, B., Roli, F.: Wild patterns: ten years after the rise of adversarial machine learning. Pattern Recognit. **84**, 317–331 (2018)

7. Boyd, S., Vandenberghe, L.: Convex Optimization. Cambridge University Press, Cambridge (2004)
8. Brendel, W., Rauber, J., Bethge, M.: Decision-based adversarial attacks: reliable attacks against black-box machine learning models. arXiv preprint arXiv:1712.04248 (2017)
9. Card, D., Zhang, M., Smith, N.A.: Deep weighted averaging classifiers. In: Proceedings of the Conference on Fairness, Accountability, and Transparency, FAT* 2019, pp. 369–378. ACM, New York (2019). https://doi.org/10.1145/3287560.3287595
10. Carlini, N.: Is Am I (attacks meet interpretability) robust to adversarial examples? arXiv preprint arXiv:1902.02322 (2019)
11. Carlini, N., et al.: On evaluating adversarial robustness. arXiv preprint arXiv:1902.06705 (2019)
12. Carlini, N., Wagner, D.: Adversarial examples are not easily detected: bypassing ten detection methods. In: Proceedings of the 10th ACM Workshop on Artificial Intelligence and Security, pp. 3–14. ACM (2017)
13. Carlini, N., Wagner, D.: Towards evaluating the robustness of neural networks. In: 2017 IEEE Symposium on Security and Privacy (SP), pp. 39–57. IEEE (2017)
14. Cheng, S., Dong, Y., Pang, T., Su, H., Zhu, J.: Improving black-box adversarial attacks with a transfer-based prior. In: Advances in Neural Information Processing Systems 32, pp. 10934–10944. Curran Associates, Inc. (2019)
15. Chollet, F., et al.: Keras (2015). https://keras.io
16. Cohen, J.M., Rosenfeld, E., Kolter, J.Z.: Certified adversarial robustness via randomized smoothing. arXiv preprint arXiv:1902.02918 (2019)
17. Cullina, D., Bhagoji, A.N., Mittal, P.: Pac-learning in the presence of adversaries. In: Advances in Neural Information Processing Systems, pp. 230–241 (2018)
18. Dalvi, N., Domingos, P., Sanghai, S., Verma, D., et al.: Adversarial classification. In: Proceedings of the tenth ACM SIGKDD International Conference on Knowledge Discovery and Data Mining, pp. 99–108. ACM (2004)
19. De Vries, H., Memisevic, R., Courville, A.C.: Deep learning vector quantization. In: ESANN (2016)
20. Deng, J., Dong, W., Socher, R., Li, L.J., Li, K., Fei-Fei, L.: ImageNet: a large-scale hierarchical image database. In: CVPR09 (2009)
21. Doob, J.L.: Regularity properties of certain families of chance variables. Trans. Am. Math. Soc. **47**(3), 455–486 (1940)
22. Ebrahimi, J., Rao, A., Lowd, D., Dou, D.: HotFlip: white-box adversarial examples for text classification. arXiv preprint arXiv:1712.06751 (2017)
23. Elson, J., Douceur, J.J., Howell, J., Saul, J.: Asirra: a CAPTCHA that exploits interest-aligned manual image categorization. In: Proceedings of 14th ACM Conference on Computer and Communications Security (CCS). Association for Computing Machinery, Inc., October 2007
24. Engstrom, L., Madry, A.: Understanding the landscape of adversarial robustness. Ph.D. thesis, Massachusetts Institute of Technology (2019)
25. Friedman, J., Hastie, T., Tibshirani, R.: The Elements of Statistical Learning, vol. 1. Springer, New York (2001). https://doi.org/10.1007/978-0-387-21606-5
26. Gal, Y.: Uncertainty in deep learning. Ph.D. thesis, University of Cambridge (2016)
27. Gal, Y., Ghahramani, Z.: Dropout as a Bayesian approximation: representing model uncertainty in deep learning. In: International Conference on Machine Learning, pp. 1050–1059 (2016)
28. Gelada, C., Buckman, J.: Bayesian neural networks need not concentrate (2020). https://jacobbuckman.com/2020-01-22-bayesian-neural-networks-need-not-concentrate/

29. Gilmer, J., Adams, R.P., Goodfellow, I., Andersen, D., Dahl, G.E.: Motivating the rules of the game for adversarial example research. arXiv preprint arXiv:1807.06732 (2018)
30. Goodfellow, I., Bengio, Y., Courville, A.: Deep Learning. MIT Press, Cambridge (2016). http://www.deeplearningbook.org
31. Goodfellow, I.J., Shlens, J., Szegedy, C.: Explaining and harnessing adversarial examples. arXiv preprint arXiv:1412.6572 (2014)
32. Gourdeau, P., Kanade, V., Kwiatkowska, M., Worrell, J.: On the hardness of robust classification. In: Advances in Neural Information Processing Systems, pp. 7444–7453 (2019)
33. Grumer, C., Peck, J., Olumofin, F., Nascimento, A., De Cock, M.: Hardening DGA classifiers utilizing IVAP. In: IEEE Big Data (2019)
34. Guo, C., Pleiss, G., Sun, Y., Weinberger, K.Q.: On calibration of modern neural networks. In: Precup, D., Teh, Y.W. (eds.) Proceedings of the 34th International Conference on Machine Learning. Proceedings of Machine Learning Research, vol. 70, pp. 1321–1330. PMLR, International Convention Centre, Sydney, Australia, 06–11 August 2017
35. Guo, Y., Yan, Z., Zhang, C.: Subspace attack: exploiting promising subspaces for query-efficient black-box attacks. In: Advances in Neural Information Processing Systems 32, pp. 3825–3834. Curran Associates, Inc. (2019)
36. He, K., Zhang, X., Ren, S., Sun, J.: Deep residual learning for image recognition. In: Proceedings of the IEEE Conference on Computer Vision and Pattern Recognition, pp. 770–778 (2016)
37. Hein, M., Andriushchenko, M., Bitterwolf, J.: Why ReLU networks yield high-confidence predictions far away from the training data and how to mitigate the problem. In: Proceedings of the IEEE Conference on Computer Vision and Pattern Recognition, pp. 41–50 (2019)
38. Hoeffding, W.: Probability inequalities for sums of bounded random variables. In: Fisher, N.I., Sen, P.K. (eds.) The Collected Works of Wassily Hoeffding. SSS, pp. 409–426. Springer, New York (1994). https://doi.org/10.1007/978-1-4612-0865-5_26
39. Jaccard, P.: The distribution of the flora in the alpine zone. New Phytol. **11**(2), 37–50 (1912)
40. Jain, H., Balasubramanian, V., Chunduri, B., Varma, M.: Slice: scalable linear extreme classifiers trained on 100 million labels for related searches. In: Proceedings of the Twelfth ACM International Conference on Web Search and Data Mining, pp. 528–536. ACM (2019)
41. Jaynes, E.T.: Prior probabilities. IEEE Trans. Syst. Sci. Cybern. **4**(3), 227–241 (1968)
42. Jordan, M.I., Ghahramani, Z., Jaakkola, T.S., Saul, L.K.: An introduction to variational methods for graphical models. Mach. Learn. **37**(2), 183–233 (1999)
43. Kanbak, C., Moosavi-Dezfooli, S.M., Frossard, P.: Geometric robustness of deep networks: analysis and improvement. In: Proceedings of the IEEE Conference on Computer Vision and Pattern Recognition, pp. 4441–4449 (2018)
44. Kannan, H., Kurakin, A., Goodfellow, I.: Adversarial logit pairing. arXiv preprint arXiv:1803.06373 (2018)
45. Kingma, D.P., Ba, J.: Adam: A method for stochastic optimization. arXiv preprint arXiv:1412.6980 (2014)
46. Klaus, B., Strimmer, K.: fdrtool: Estimation of (Local) False Discovery Rates and Higher Criticism (2015). https://CRAN.R-project.org/package=fdrtool. r package version 1.2.15

47. Krizhevsky, A., Hinton, G.: Learning multiple layers of features from tiny images. Technical Report, Citeseer (2009)
48. LeCun, Y., Bottou, L., Bengio, Y., Haffner, P., et al.: Gradient-based learning applied to document recognition. Proc. IEEE **86**(11), 2278–2324 (1998)
49. Lowd, D., Meek, C.: Adversarial learning. In: Proceedings of the eleventh ACM SIGKDD International Conference on Knowledge Discovery in Data Mining, pp. 641–647. ACM (2005)
50. Maddox, W.J., Izmailov, P., Garipov, T., Vetrov, D.P., Wilson, A.G.: A simple baseline for Bayesian uncertainty in deep learning. In: Advances in Neural Information Processing Systems 32, pp. 13132–13143. Curran Associates, Inc. (2019)
51. Madry, A., Makelov, A., Schmidt, L., Tsipras, D., Vladu, A.: Towards deep learning models resistant to adversarial attacks. arXiv preprint arXiv:1706.06083 (2017)
52. Manokhin, V.: Multi-class probabilistic classification using inductive and cross Venn-Abers predictors. In: Conformal and Probabilistic Prediction and Applications, pp. 228–240 (2017)
53. McDiarmid, C.: On the method of bounded differences. Surv. Comb. **141**(1), 148–188 (1989)
54. Meinke, A., Hein, M.: Towards neural networks that provably know when they don't know. arXiv preprint arXiv:1909.12180 (2019)
55. Moosavi-Dezfooli, S.M., Fawzi, A., Fawzi, O., Frossard, P.: Universal adversarial perturbations. In: Proceedings of the IEEE Conference on Computer Vision and Pattern Recognition, pp. 1765–1773 (2017)
56. Moosavi-Dezfooli, S.M., Fawzi, A., Frossard, P.: DeepFool: a simple and accurate method to fool deep neural networks. In: Proceedings of the IEEE Conference on Computer Vision and Pattern Recognition, pp. 2574–2582 (2016)
57. Murphy, K.P.: Machine Learning: A Probabilistic Perspective. MIT Press, Cambridge (2012)
58. Nakkiran, P.: Adversarial robustness may be at odds with simplicity. arXiv preprint arXiv:1901.00532 (2019)
59. Narodytska, N., Kasiviswanathan, S.P.: Simple black-box adversarial perturbations for deep networks. arXiv preprint arXiv:1612.06299 (2016)
60. Netzer, Y., Wang, T., Coates, A., Bissacco, A., Wu, B., Ng, A.Y.: Reading digits in natural images with unsupervised feature learning. In: NIPS Workshop on Deep Learning and Unsupervised Feature Learning (2011)
61. Papernot, N., McDaniel, P.: Deep k-nearest neighbors: towards confident, interpretable and robust deep learning. arXiv preprint arXiv:1803.04765 (2018)
62. Peck, J., Goossens, B., Saeys, Y.: Detecting adversarial examples with inductive Venn-Abers predictors. In: European Symposium on Artificial Neural Networks, Computational Intelligence and Machine Learning, pp. 143–148 (2019)
63. Platt, J., et al.: Probabilistic outputs for support vector machines and comparisons to regularized likelihood methods. Adv. Large Margin Cl. **10**(3), 61–74 (1999)
64. Price, D., Knerr, S., Personnaz, L., Dreyfus, G.: Pairwise neural network classifiers with probabilistic outputs. In: Advances in Neural Information Processing Systems, pp. 1109–1116 (1995)
65. Qin, Y., Carlini, N., Cottrell, G., Goodfellow, I., Raffel, C.: Imperceptible, robust, and targeted adversarial examples for automatic speech recognition. In: Chaudhuri, K., Salakhutdinov, R. (eds.) Proceedings of the 36th International Conference on Machine Learning. Proceedings of Machine Learning Research, Long Beach, California, USA, vol. 97, pp. 5231–5240. PMLR, June 2019

66. R Core Team: R: A Language and Environment for Statistical Computing. R Foundation for Statistical Computing, Vienna, Austria (2015). https://www.R-project.org/

67. Raghunathan, A., Steinhardt, J., Liang, P.S.: Semidefinite relaxations for certifying robustness to adversarial examples. In: Advances in Neural Information Processing Systems, pp. 10877–10887 (2018)

68. Rauber, J., Brendel, W., Bethge, M.: Foolbox v0. 8.0: A Python toolbox to benchmark the robustness of machine learning models. arXiv preprint arXiv:1707.04131 5 (2017)

69. Schmidt, L., Santurkar, S., Tsipras, D., Talwar, K., Madry, A.: Adversarially robust generalization requires more data. In: Advances in Neural Information Processing Systems, pp. 5014–5026 (2018)

70. Schneier, B.: Schneier's law (2011). https://www.schneier.com/blog/archives/2011/04/schneiers_law.html

71. Shafahi, A., et al.: Adversarial training for free! In: Advances in Neural Information Processing Systems 32, pp. 3353–3364. Curran Associates, Inc. (2019)

72. Shafer, G., Vovk, V.: A tutorial on conformal prediction. J. Mach. Learn. Res. **9**, 371–421 (2008)

73. Shen, J., et al.: Lingvo: a modular and scalable framework for sequence-to-sequence modeling. arXiv preprint arXiv:1902.08295 (2019)

74. Sinha, A., Namkoong, H., Duchi, J.: Certifiable distributional robustness with principled adversarial training. In: International Conference on Learning Representations (2018). https://openreview.net/forum?id=Hk6kPgZA-

75. Sitawarin, C., Wagner, D.: On the robustness of deep k-nearest neighbors. arXiv preprint arXiv:1903.08333 (2019)

76. So, D., Le, Q., Liang, C.: The evolved transformer. In: Chaudhuri, K., Salakhutdinov, R. (eds.) Proceedings of the 36th International Conference on Machine Learning. Proceedings of Machine Learning Research, Long Beach, California, USA, vol. 97, pp. 5877–5886. PMLR, June 2019

77. Su, J., Vargas, D.V., Sakurai, K.: One pixel attack for fooling deep neural networks. IEEE Trans. Evol. Comput. **23**, 828–841 (2019)

78. Sun, R.: Optimization for deep learning: theory and algorithms. arXiv preprint arXiv:1912.08957 (2019)

79. Szegedy, C., Vanhoucke, V., Ioffe, S., Shlens, J., Wojna, Z.: Rethinking the inception architecture for computer vision. In: Proceedings of the IEEE Conference on Computer Vision and Pattern Recognition, pp. 2818–2826 (2016)

80. Szegedy, C., et al.: Intriguing properties of neural networks. arXiv preprint arXiv:1312.6199 (2013)

81. Tanay, T., Griffin, L.: A boundary tilting persepective on the phenomenon of adversarial examples. arXiv preprint arXiv:1608.07690 (2016)

82. Toccaceli, P.: Venn-ABERS predictor (2017). https://github.com/ptocca/VennABERS

83. Tuy, H.: Convex Analysis and Global Optimization. Springer, New York (1998). https://doi.org/10.1007/978-1-4757-2809-5

84. Vorobeychik, Y., Li, B.: Optimal randomized classification in adversarial settings. In: Proceedings of the 2014 International Conference on Autonomous Agents and Multi-Agent Systems, pp. 485–492. International Foundation for Autonomous Agents and Multiagent Systems (2014)

85. Vovk, V., Gammerman, A., Shafer, G.: Algorithmic Learning in a Random World. Springer, Boston (2005). https://doi.org/10.1007/b106715

86. Vovk, V., Petej, I., Fedorova, V.: Large-scale probabilistic predictors with and without guarantees of validity. In: Advances in Neural Information Processing Systems, pp. 892–900 (2015)
87. Wasserman, L.: Frasian inference. Stat. Sci., 322–325 (2011)
88. Xiao, H., Rasul, K., Vollgraf, R.: Fashion-MNIST: a novel image dataset for benchmarking machine learning algorithms (2017)
89. Yin, D., Ramchandran, K., Bartlett, P.: Rademacher complexity for adversarially robust generalization. arXiv preprint arXiv:1810.11914 (2018)
90. Zagoruyko, S., Komodakis, N.: Wide residual networks. arXiv preprint arXiv:1605.07146 (2016)
91. Zhang, C., Bengio, S., Hardt, M., Recht, B., Vinyals, O.: Understanding deep learning requires rethinking generalization. arXiv preprint arXiv:1611.03530 (2016)

Machine Learning Methods for Ordinal Classification with Additional Relative Information

Mengzi Tang[1(✉)], Raúl Pérez-Fernández[1,2], and Bernard De Baets[1]

[1] KERMIT, Department of Data Analysis and Mathematical Modelling,
Ghent University, Coupure Links 653, 9000 Ghent, Belgium
{mengzi.tang,bernard.debaets}@ugent.be
[2] UNIMODE, Department of Statistics and O.R. and Mathematics Didactics,
University of Oviedo, C/ Federico García Lorca 18, 3307 Oviedo, Spain
perezfernandez@uniovi.es

Abstract. The ability to learn an accurate classification process is often limited by the amount of labeled data. Incorporating additional information into the learning process for overcoming this limitation has been a popular research topic. In this work, we focus on ordinal classification problems that are provided with limited absolute information and additional relative information. We modify some classical machine learning methods to combine both types of information. Moreover, we propose a new distance metric learning method to exploit both types of information for learning a suitable distance metric that can be incorporated into the augmented method of k nearest neighbors for ordinal classification. The experimental results show that our method is competitive with other modified machine learning methods and considering additional relative information leads to a better performance.

Keywords: Machine learning methods · Ordinal classification · Distance metric learning · Absolute information · Relative information

1 Introduction

Ordinal classification problems naturally appear in many research areas, such as medical research [5] and social sciences [7]. For instance, on the online shopping website Amazon, customers use ordinal labels to evaluate products according to their preferences. The ordinal scale could be represented by four labels such as "bad", "average", "good" and "excellent". Machine learning methods deal with large manually-labeled datasets for solving these problems. However, the performance of these methods is usually limited by the amount of absolute information (i.e., examples with given class labels). Moreover, it is usually time-consuming and costly to collect a large amount of absolute information. Fortunately, gathering a large amount of additional relative information (i.e., preference orders for couples of examples) is much easier. Even though specific class labels from

© Springer Nature Switzerland AG 2020
B. Bogaerts et al. (Eds.): BNAIC 2019/BENELEARN 2019, CCIS 1196, pp. 126–136, 2020.
https://doi.org/10.1007/978-3-030-65154-1_7

absolute information are much more informative than preference orders from relative information, in order to demonstrate the usefulness of relative information, the main challenge of this work is to combine a small amount of absolute information and a large amount of relative information for ordinal classification.

There are many research works that have discussed and validated the importance of considering additional information [12]. For example, in the field of soft-label classification [13,14,19], the additional information consists of probability scores. A basic assumption is that there is some uncertainty associated with the class label that should be associated with each of the examples [12]. Experts are invited to assign class labels to examples and provide the corresponding additional information that reflects how certain they are on the class label that should be specified. A natural way of solving the soft-label classification problem is to modify classical machine learning methods, e.g.., logistic regression and support vector machines. Inspired by this idea, for our problem setting in which there is a small amount of absolute information and a large amount of relative information, we use a similar idea to modify classical machine learning methods by combining both types of information.

Furthermore, there are many related methods that have been proposed for solving ordinal classification problems. In addition to some classical methods [8], such as naive methods, ordinal binary decomposition methods and threshold methods, distance metric learning methods [2] are also popular. For example, Fouad et al. [6] incorporated additional information into ordinal classification tasks by changing the default distance metric in the input space based on a natural order between class labels. Their experimental results show that the proposed ordinal-based distance metric learning improves the ordinal classification performance. Recently, Nguyen et al. [11] considered ordinal information as local constraints and proposed a method that incorporates these constraints into a distance metric learning task. Their experiments demonstrated that considering the order of class labels could lead to learning a better distance metric.

In this work, firstly, we modify some classical machine learning methods by incorporating additional relative information. Subsequently, similarly to the assumption that close examples are assumed to have the same class label, we assume that close couples tend to have the same order relation, an assumption that has been successfully validated in previous work [16]. Inspired by the work of Nguyen et al. [11], we learn a better distance metric by imposing different distance constraints for absolute and relative information, and incorporate the learned distance metric into the augmented method of k-NN [16] for ordinal classification. Finally, we test the modified classical machine learning methods and the distance metric learning methods on some benchmark datasets. The experiments show the benefits of considering additional relative information.

2   Problem Description

Formally, the input data includes two types of information: absolute information and relative information. The first type of information is collected in a set $\mathcal{A} = \{(\mathbf{x}_1, y_1), (\mathbf{x}_2, y_2), ..., (\mathbf{x}_n, y_n)\}$ with a set of input examples $\mathcal{D} = \{\mathbf{x}_1, \mathbf{x}_2, ..., \mathbf{x}_n\}$, where $\mathbf{x}_i = (x_{i1}, ..., x_{id})$ belong to the input space $\mathcal{X} \subseteq \mathbb{R}^d$ and the class labels y_i belong to the output space $\mathcal{Y} = \{C_1, C_2, ..., C_r\}$. The class labels are assumed to be ordered as follows $C_1 \prec C_2 \prec ... \prec C_r$. The second type of information is collected in a set $\mathcal{R} = \{((\mathbf{a}^1, \mathbf{b}^1), R_1), ..., ((\mathbf{a}^m, \mathbf{b}^m), R_m)\}$. We denote the set of couples for which we have relative information by $\mathcal{C} = \{(\mathbf{a}^1, \mathbf{b}^1), (\mathbf{a}^2, \mathbf{b}^2), ..., (\mathbf{a}^m, \mathbf{b}^m)\}$. It is assumed that, if a couple $(\mathbf{a}, \mathbf{b})$ belongs to $\mathcal{C}$, then also the couple $(\mathbf{b}, \mathbf{a})$ belongs to $\mathcal{C}$. For any $p \in \{1, ..., m\}$, an order relation $R_p = \prec$ indicates that $\mathbf{b}^p$ is preferred to $\mathbf{a}^p$ and an order relation $R_p = \succ$ indicates that $\mathbf{a}^p$ is preferred to $\mathbf{b}^p$. Note that here we impose that both $\mathbf{a}^p$ and $\mathbf{b}^p$ may not be equally preferred. Additionally, it is assumed that whenever $((\mathbf{a}, \mathbf{b}), R) \in \mathcal{R}$, it also holds that $((\mathbf{b}, \mathbf{a}), R^\mathrm{T}) \in \mathcal{R}$, where R^T represents the transpose of R. A main characteristic of our problem is that the amount of absolute information is typically smaller than the amount of relative information, i.e., $n \ll m$.

3   Machine Learning Methods with Additional Relative Information

In this section, firstly, we modify some classical model-based machine learning methods to incorporate additional relative information. Secondly, we extend an instance-based distance metric learning method to combine absolute and relative information.

3.1   Classical Machine Learning Methods

Proportional Odds Model [10] **with Additional Relative Information.** The goal of the proportional odds model (POM) is to use a logistic function to predict the probabilities of the different possible outputs. Formally, the cumulative probability is modeled as the logistic function as follows:

$$P(y_i \le j | \mathbf{x}_i) = \phi(\theta_j - \mathbf{w} \cdot \mathbf{x}_i) = \frac{1}{1 + \exp(\mathbf{w} \cdot \mathbf{x}_i - \theta_j)}, \tag{1}$$

where the vector $\mathbf{w}$ is common across all class labels, the vector of thresholds θ is used for separating different class labels and $\phi(t) = \frac{1}{1+\exp(-t)}$. The class label C_k is represented as the interval $C_k \in [\theta_{k-1}, \theta_k]$, where $\theta_0 = -\infty$ and $\theta_r = +\infty$. The loss function is defined as the negative log-likelihood as follows:

$$\mathcal{L}(\mathbf{w}, \theta) = -\sum_{i=1}^{n} \log(\phi(\theta_{y_i} - \mathbf{w} \cdot \mathbf{x}_i) - \phi(\theta_{y_i-1} - \mathbf{w} \cdot \mathbf{x}_i)). \tag{2}$$

In our problem setting, additional relative information is provided. For each ordinal couple $(\mathbf{a}^j, \mathbf{b}^j)$ with an order relation $\succ$, we get the inequalities $\mathbf{w} \cdot \mathbf{a}^j > \mathbf{w} \cdot \mathbf{b}^j$. In order to incorporate a large margin that separates an example from another one that is regarded as worse [18], the corresponding constraints are formulated as follows:

$$\mathbf{w} \cdot \mathbf{a}^j - \mathbf{w} \cdot \mathbf{b}^j \ge 1. \tag{3}$$

By considering a soft margin to deal with some non-separable cases, we define the new objective function as follows:

$$\min_{\mathbf{w},\theta} \quad \mathcal{L}(\mathbf{w},\theta) + \frac{\alpha}{m}\sum_{j=1}^{m}\eta_j + \frac{\lambda}{2}\|\mathbf{w}\|_2^2$$

$$\text{s.t.} \quad \mathbf{w}\cdot\mathbf{a}^j - \mathbf{w}\cdot\mathbf{b}^j \geq 1 - \eta_j, \quad (\text{case } R_j = \succ), \tag{4}$$

$$\eta_j \geq 0, \quad \forall j,$$

where m is the number of constraints, α is a parameter to control the impact from relative information, $\|\mathbf{w}\|_2$ is the regularizer to avoid overfitting, λ is the regularization parameter and the η_j are slack variables. We refer to the proportional odds model with additional relative information as POM-R.

Support Vector Learning for Ordinal Regression [9] **with Additional Relative Information.** The goal of support vector learning for ordinal regression (SVOR) is to consider a utility function $U(\mathbf{x}) = \mathbf{w}\cdot\mathbf{x}$ that is related to a mapping h from objects to ranks by $h(\mathbf{x}) = C_k \Leftrightarrow U(\mathbf{x}) \in [\theta_{k-1}, \theta_k]$, where $\theta_0 = -\infty$ and $\theta_r = +\infty$. The objective function is formulated as follows:

$$\min_{\mathbf{w},\theta} \quad \frac{1}{2}\|\mathbf{w}\|_2^2 + C\sum_{i=1}^{n}\sum_{j\neq i}\xi_{i,j}$$

$$\text{s.t.} \quad z_{i,j}(\mathbf{w}\cdot\mathbf{x}_i - \mathbf{w}\cdot\mathbf{x}_j) \geq 1 - \xi_{i,j}, \tag{5}$$

$$\xi_{i,j} \geq 0, \quad \forall i,j,$$

where $C > 0$ is a trade-off parameter, $z_{i,j} = +1$ when $y_i \succ y_j$ and $z_{i,j} = -1$ when $y_i \prec y_j$. The rank boundaries θ_k are estimated as

$$\theta_k = \frac{U(\mathbf{x}_1;\mathbf{w}^*) + U(\mathbf{x}_2;\mathbf{w}^*)}{2}, \tag{6}$$

where $\mathbf{w}^*$ is the optimal weight vector, $(\mathbf{x}_1,\mathbf{x}_2) = \arg\min_{(i,j),y_i=C_{k+1},y_j=C_k}[U(\mathbf{x}_i;\mathbf{w}^*) - U(\mathbf{x}_j;\mathbf{w}^*)]$, which means that the optimal thresholds θ_k for the class label C_k lie in the middle of the utilities of the closest examples of the class labels C_k and C_{k+1}.

Here, for additional relative information, we incorporate the same constraints as in Eq. (4). The new objective function is formulated as follows:

$$\min_{\mathbf{w},\theta} \quad \frac{1}{2}\|\mathbf{w}\|_2^2 + C\sum_{i=1}^{n}\sum_{j\neq i}\xi_{i,j} + \frac{\alpha}{m}\sum_{j=1}^{m}\eta_j$$

$$\text{s.t.} \quad z_{i,j}(\mathbf{w}\cdot\mathbf{x}_i - \mathbf{w}\cdot\mathbf{x}_j) \geq 1 - \xi_{i,j},$$

$$\xi_{i,j} \geq 0, \quad \forall i,j, \tag{7}$$

$$\mathbf{w}\cdot\mathbf{a}^j - \mathbf{w}\cdot\mathbf{b}^j \geq 1 - \eta_j, \quad (\text{case } R_j = \succ),$$

$$\eta_j \geq 0, \quad \forall j.$$

We refer to the support vector learning for ordinal regression with additional relative information as SVOR-R.

Support Vector Ordinal Regression [4] with Additional Relative Information. The goal of support vector ordinal regression is to find an optimal mapping direction $\mathbf{w}$ and $r - 1$ thresholds, which define $r - 1$ parallel discriminant hyperplanes for the r class labels. Differently to the above-mentioned SVOR, here the methods use two different ways to estimate empirical errors by considering thresholds. One way is to consider the explicit constraints on thresholds (this method is referred to as SVOREX). For each threshold θ_j, the empirical errors are computed for the examples from the two adjacent classes C_j and C_{j+1}. The objective function is formulated as follows:

$$\min_{\mathbf{w},\theta} \quad \frac{1}{2} \|\mathbf{w}\|_2^2 + C \sum_{j=1}^{r-1} \left(\sum_{i=1}^{n^j} \xi_i^j + \sum_{i=1}^{n^{j+1}} \xi_i^{*j+1} \right)$$

$$\text{s.t.} \quad \mathbf{w} \cdot \mathbf{x}_i^j - \theta_j \leq -1 + \xi_i^j \ ,$$

$$\xi_i^j \geq 0, \quad \forall i = 1, ..., n^j \ , \tag{8}$$

$$\mathbf{w} \cdot \mathbf{x}_i^{j+1} - \theta_j \geq +1 - \xi_i^{*j+1} \ ,$$

$$\xi_i^{*j+1} \geq 0, \quad \forall i = 1, ..., n^{j+1} \ ,$$

where $\theta_{j-1} \leq \theta_j$, for $j = 2, ..., r - 1$, the n^j is the number of examples with the class label C_j and ξ_i^j, ξ_i^{*j+1} are slack variables.

The other way is to consider the implicit constraints on thresholds (this method is referred to as SVORIM). The examples in all the classes are incorporated to estimate the errors for all thresholds. The objective function is formulated as follows:

$$\min_{\mathbf{w},\theta} \quad \frac{1}{2} \|\mathbf{w}\|_2^2 + C \sum_{j=1}^{r-1} \left(\sum_{k=1}^{j} \sum_{i=1}^{n^k} \xi_{ki}^j + \sum_{k=j+1}^{r} \sum_{i=1}^{n^r} \xi_{ki}^{*j} \right)$$

$$\text{s.t.} \quad \mathbf{w} \cdot \mathbf{x}_i^k - \theta_j \leq -1 + \xi_{ki}^j, \ \xi_{ki}^j \geq 0 \ ,$$

$$\forall k = 1, ..., j, \quad i = 1, ..., n^k \ , \tag{9}$$

$$\mathbf{w} \cdot \mathbf{x}_i^k - \theta_j \geq +1 - \xi_{ki}^{*j}, \ \xi_{ki}^{*j} \geq 0 \ ,$$

$$\forall k = j + 1, ..., r, \quad i = 1, ..., n^k \ .$$

Here, similarly to the above-mentioned modified methods, we incorporate additional relative information into the two methods by setting the same constraints as in Eq. (4). The new objective functions are omitted due to the space limit. We refer to the corresponding modified methods as SVOREX-R and SVORIM-R, respectively.

3.2   Distance Metric Learning Methods

In this paper, we extend the idea of Nguyen et al. [11] to process absolute and relative information. For absolute information, if the neighbor example has the same class label as the given example, the distance to the given example is expected to be small. Otherwise, the distance to the given example is expected to be large. Moreover, the examples with different class labels are expected to be separated according to the differences between their class labels. The corresponding constraints $\mathcal{R} = \mathcal{R}_1 \cup \mathcal{R}_2$ are described as:

$$\mathcal{R}_1 = \{(i,j,l) \mid i,j,l \in \{1,...,n\}, \mathbf{x}_j, \mathbf{x}_l \in \mathcal{N}(\mathbf{x}_i), y_i = y_j \neq y_l\}, \quad (10)$$

$$\mathcal{R}_2 = \{(i,j,l) \mid i,j,l \in \{1,...,n\}, \mathbf{x}_j, \mathbf{x}_l \in \mathcal{N}(\mathbf{x}_i), (y_i \succ y_j \succ y_l) \vee (y_i \prec y_j \prec y_l)\}, \quad (11)$$

where $\mathcal{N}(\mathbf{x}_i)$ is the neighborhood of $\mathbf{x}_i$ containing the nearest neighbor examples. In the following experiments, we set the number of nearest neighbor examples as 5.

We refer to the distance metric learning method as DMLA, solving the following optimization problem:

$$
\begin{aligned}
\min_{\mathbf{M},\xi} \quad & f_L(\mathbf{M}) = \lambda \mathrm{Tr}(\mathbf{M}) + \frac{1}{C} \sum_{(i,j,l) \in \mathcal{R}} \xi_{ijl} \\
\text{s.t.} \quad & d_{\mathbf{M}}^2(\mathbf{x}_i, \mathbf{x}_l) - d_{\mathbf{M}}^2(\mathbf{x}_i, \mathbf{x}_j) \geq 1 - \xi_{ijl} \\
& \xi_{ijl} \geq 0, \quad \forall (i,j,l) \in \mathcal{R} \\
& \mathbf{M} \succeq 0,
\end{aligned}
\qquad (12)
$$

where λ is the regularization parameter, $\mathrm{Tr}(\mathbf{M})$ is the trace of the matrix $\mathbf{M}$, which is computed as the sum of all diagonal elements, C is the number of constraints in $\mathcal{R}$, $d_{\mathbf{M}}^2(\mathbf{x}_i, \mathbf{x}_j) = (\mathbf{x}_i - \mathbf{x}_j)^{\mathrm{T}} \mathbf{M}(\mathbf{x}_i - \mathbf{x}_j)$ and ξ_{ijl} are slack variables.

However, for relative information, there are no given class labels and we cannot explicitly use the above distance constraints. Here, we assume that close couples have the same direction and then consider relative information by setting different distance constraints:

$$\mathcal{R}' = \{(p,q,t) \mid p,q,t \in \{1,...,m\}, (\mathbf{a}^q, \mathbf{b}^q), (\mathbf{a}^t, \mathbf{b}^t) \in \mathcal{N}((\mathbf{a}^p, \mathbf{b}^p)), \zeta^p = \zeta^q \neq \zeta^t\}, \quad (13)$$

where ζ represents the order relation, $\zeta \in \{\succ, \prec\}$ and $\mathcal{N}((\mathbf{a}^p, \mathbf{b}^p))$ is the neighborhood of the couple $(\mathbf{a}^p, \mathbf{b}^p)$. Here, we compute the distance between couples by $d_*((\mathbf{u}, \mathbf{v}), (\mathbf{w}, \mathbf{t})) = d(\mathbf{u}, \mathbf{w}) + d(\mathbf{v}, \mathbf{t})$, where $d(\mathbf{u}, \mathbf{v}) = \sqrt{\sum_{i=1}^{d}(u_i - v_i)^2}$.

We refer to our proposed distance metric learning method as DMLAR, solving the following problem:

$$\min_{\mathbf{M},\xi,\eta} \quad f_L(\mathbf{M}) = \lambda \mathrm{Tr}(\mathbf{M}) + \frac{\alpha}{C} \sum_{(i,j,l)\in\mathcal{R}} \xi_{ijl} + \frac{\beta}{D} \sum_{(p,q,t)\in\mathcal{R}'} \eta_{pqt}$$

$$\text{s.t.} \quad d_{\mathbf{M}}^2(\mathbf{x}_i, \mathbf{x}_l) - d_{\mathbf{M}}^2(\mathbf{x}_i, \mathbf{x}_j) \geq 1 - \xi_{ijl}$$

$$d_{*\mathbf{M}}^2((\mathbf{a}^p, \mathbf{b}^p), (\mathbf{a}^t, \mathbf{b}^t)) - d_{*\mathbf{M}}^2((\mathbf{a}^p, \mathbf{b}^p), (\mathbf{a}^q, \mathbf{b}^q)) \geq 1 - \eta_{pqt} \qquad (14)$$

$$\xi_{ijl} \geq 0, \quad \forall (i,j,l) \in \mathcal{R}$$

$$\eta_{pqt} \geq 0, \quad \forall (p,q,t) \in \mathcal{R}'$$

$$\mathbf{M} \succcurlyeq 0 \,,$$

where α is the parameter to control the impact from absolute information, β is the parameter to control the impact from relative information, D is the number of constraints in $\mathcal{R}'$, $d_{*\mathbf{M}}((\mathbf{a}^s, \mathbf{b}^s), (\mathbf{a}^r, \mathbf{b}^r)) = d_{\mathbf{M}}(\mathbf{a}^s, \mathbf{a}^r) + d_{\mathbf{M}}(\mathbf{b}^s, \mathbf{b}^r)$ and η_{pqt} are slack variables. Note that, in order to reduce the computational complexity, we replace $d_{*\mathbf{M}}^2((\mathbf{a}^s, \mathbf{b}^s), (\mathbf{a}^r, \mathbf{b}^r))$ by $d_{\mathbf{M}}^2(\mathbf{a}^s, \mathbf{a}^r) + d_{\mathbf{M}}^2(\mathbf{b}^s, \mathbf{b}^r)$, where the additional term obtained in the expansion of the square of the sum of both distances is ignored for avoiding computing the gradient of square roots. We use subgradient descent methods to update $\mathbf{M}$ until the above objective function converges.

Ultimately, the learned distance metric is used within k-NN. We refer to this method as DMLA k-NN or DMLAR k-NN (depending on whether only absolute or both absolute and relative information is used for learning the distance metric).

4   Experiments

4.1   Datasets

We perform our experiments on some datasets from real ordinal classification problems and other datasets from discretized regression problems. The former datasets are from some open repositories, UCI machine learning repository [1] and mldata.org [15]. The latter datasets are provided by Chu [3]. Table 1 describes the characteristics of these datasets. We use ten-fold cross-validation to get the performance.

Note that these datasets do not contain relative information. Due to the difficulties of getting relative information from real datasets, we simulate synthetic data by generating relative information from the given absolute information. For each dataset, one of the folds is used for testing. The other folds are used for collecting absolute and relative information. We randomly select 5% of the other folds as absolute information and the remaining 95% for generating relative information by transforming the class labels into order relations between examples. More in detail, if there are two training examples and their class labels are $y_1 = C_1$ and $y_2 = C_2$ with the order relation $C_1 \prec C_2$, we will recreate two couples $(\mathbf{a}^1, \mathbf{b}^1)$ with $\mathbf{a}^1 \prec \mathbf{b}^1$ and $(\mathbf{b}^1, \mathbf{a}^1)$ with $\mathbf{b}^1 \succ \mathbf{a}^1$. Note that, if two examples have the same class label, no couple is generated. For more details on the generation process for relative information, we refer to our previous work [16].

Table 1. Description of the benchmark datasets.

Dataset	#Examples	#Features	#Classes
Real ordinal classification datasets			
Toy (TO)	300	2	5
Balance-scale (BS)	625	4	3
Eucalyptus (EU)	736	91	5
Swd (SW)	1000	10	4
Lev (LE)	1000	4	5
Winequality-red (WR)	1599	11	6
Car (CA)	1728	21	4
Discretized regression datasets			
Housing5 (HO5)	506	14	5
Abalone5 (AB5)	4177	11	5
Bank1-5 (BA1-5)	8192	8	5
Bank2-5 (BA2-5)	8192	32	5
Computer1-5 (CO1-5)	8192	12	5
Computer2-5 (CO2-5)	8192	21	5
Housing10 (HO10)	506	14	10
Abalone10 (AB10)	4177	11	10
Bank1-10 (BA1-10)	8192	8	10
Bank2-10 (BA2-10)	8192	32	10
Computer1-10 (CO1-10)	8192	12	10
Computer2-10 (CO2-10)	8192	21	10

4.2   Performance Measure

Here, as performance measure we use the C-index, which is computed as the proportion of the number of concordant pairs to the number of comparable pairs (see [17], page 50):

$$\text{C-index} = \frac{1}{\sum_{C_p \prec C_q} T_{C_p} T_{C_q}} \sum_{y_i \prec y_j} \left(\delta(y_i^* \prec y_j^*) + \frac{1}{2}\delta(y_i^* = y_j^*) \right),$$

where T_{C_p} and T_{C_q} are respectively the numbers of test examples with the class labels C_p and C_q, $\{y_i, y_j\}$ is the real ordinal pair from test examples, while $\{y_i^*, y_j^*\}$ is the corresponding predicted ordinal pair.

4.3   Performance Analysis

For each original dataset, we generate the new constructed data with only absolute information or with both absolute and relative information. We employ

Table 2. C-index for only absolute information or for both absolute and relative information from each original dataset. The best results are highlighted in boldface. The column Num represents the number of modified methods that outperform the corresponding methods. The row Num represents the number of original datasets where the modified method outperforms the corresponding method.

Dataset	For only absolute information					For absolute and relative information					Num
	POM	SVOR	SVOREX	SVORIM	DMLA k-NN	POM-R	SVOR-R	SVOREX-R	SVORIM-R	DMLAR k-NN	
TO	0.5068	0.5039	0.5192	0.4933	**0.6230**	0.5372	0.5327	0.5142	0.4959	**0.7822**	4
BS	0.8964	0.8840	0.6281	**0.8997**	0.8943	0.9283	0.9292	**0.9457**	0.9339	0.9209	5
EU	0.6895	0.7288	0.7103	0.7199	**0.7295**	0.8277	**0.8645**	0.7444	0.7789	0.8163	5
SW	0.5757	**0.7117**	0.6027	0.6385	0.6565	0.6020	**0.7521**	0.6796	0.6652	0.6979	5
LE	0.7738	**0.7962**	0.6276	0.7958	0.7593	0.7631	**0.8144**	0.8028	0.7918	0.7860	3
WR	0.5809	**0.7272**	0.7171	0.7058	0.6751	0.5809	**0.7301**	0.7111	0.6826	0.7173	2
CA	0.8287	0.8567	**0.9142**	0.9109	0.8965	0.8278	**0.9462**	0.9235	0.9201	0.9397	4
HO5	0.8256	0.7875	0.8217	**0.8467**	0.8313	0.8453	0.8405	0.8299	0.8424	**0.8642**	4
AB5	0.7913	0.6843	0.7755	**0.7916**	0.7226	**0.7980**	0.7770	0.7821	0.7896	0.7178	3
BA1-5	0.9341	0.7161	**0.9477**	**0.9477**	0.9323	0.9395	0.9446	**0.9473**	0.9472	0.9264	2
BA2-5	0.7962	0.6048	0.7966	**0.8044**	0.7382	**0.8147**	0.7970	0.8055	0.8071	0.7342	4
CO1-5	**0.8966**	0.7905	0.8938	0.8951	0.8772	0.8951	0.8821	0.8911	**0.8961**	0.8835	3
CO2-5	0.9086	0.8663	0.9095	**0.9113**	0.9049	0.9072	0.8997	0.9085	**0.9109**	0.9061	2
HO10	**0.8126**	0.7750	0.7954	0.7752	0.7817	0.7976	0.8250	0.8229	0.8191	**0.8381**	4
AB10	**0.7884**	0.6825	0.7623	0.7839	0.7170	**0.7897**	0.7826	0.7750	0.7841	0.7009	4
BA1-10	0.9446	0.6985	0.9493	**0.9497**	0.9359	0.9460	0.9446	0.9493	**0.9497**	0.9254	2
BA2-10	0.7940	0.6171	0.7895	**0.7988**	0.7294	0.8003	0.7965	0.7921	**0.8007**	0.7246	4
CO1-10	**0.8969**	0.7902	0.8922	0.8949	0.8726	**0.8989**	0.8749	0.8887	0.8959	0.8756	4
CO2-10	**0.9186**	0.8504	0.9111	0.9137	0.8970	**0.9205**	0.8913	0.8109	0.9140	0.9009	4
Num						13	19	11	12	13	

the classical machine learning methods POM, SVOR, SVOREX, SVORIM and the distance metric learning method DMLA k-NN for only absolute information from each original dataset. We employ the modified machine learning methods POM-R, SVOR-R, SVOREX-R, SVORIM-R and our proposed distance metric learning method DMLAR k-NN for both absolute and relative information from each original dataset. Their performances are shown in Table 2. In the classical and modified machine learning methods, threefold cross-validation is used to determine the parameters α, λ, C. In the distance metric learning methods, we set $\lambda = 10^{-4}$ and $\alpha = \beta = 1$.

These experimental results show that the distance metric learning methods are competitive with other classical or modified machine learning methods. Different methods get the best performance on different data, which is reasonable, because there is no method that can beat all other methods. Note that on most of the datasets at least 3 out of 5 modified methods for both absolute and relative information perform better than the corresponding methods for only absolute information. Additionally, each modified method for both absolute and relative information performs better than the corresponding method for only absolute information on at least 11 out of 19 datasets. The results show that considering additional relative information improves the performance of different machine learning methods.

The experimental results also show other phenomena. First, SVOR-R outperforms SVOR on all datasets but other modified machine learning methods only outperform their original counterparts on some datasets. For instance, SVOREX-R outperforms SVOREX only on 11 datasets. Second, when considering absolute information only, SVORIM gets a better performance than other original machine learning methods on most of the datasets.

5   Conclusion

We have modified some classical machine learning methods for ordinal classification for the setting in which limited absolute information and substantial additional relative information is available. Furthermore, we have proposed a distance metric learning method that combines absolute and relative information to satisfy different distance constraints for learning a better distance metric. This learned distance metric is incorporated into the augmented method of k-NN [16] for ordinal classification. We perform experiments on some benchmark datasets. The experimental results show that the distance metric learning method is competitive with respect to other machine learning methods and that incorporating additional relative information leads to a better performance.

References

1. Asuncion, A., Newman, D.J.: UCI machine learning repository (2007). http://www.ics.uci.edu/~mlearn/MLRepository.html
2. Bellet, A., Habrard, A., Sebban, M.: Metric learning. Synth. Lect. Artif. Intell. Mach. Learn. **9**(1), 1–151 (2015)
3. Chu, W., Ghahramani, Z.: Gaussian processes for ordinal regression. J. Mach. Learn. Res. **6**(7), 1019–1041 (2005)
4. Chu, W., Keerthi, S.S.: Support vector ordinal regression. Neural Comput. **19**(3), 792–815 (2007)
5. Doyle, O.M., et al.: Predicting progression of alzheimer's disease using ordinal regression. Plos One **9**(8), e105542 (2014)
6. Fouad, S., Tino, P.: Ordinal-based metric learning for learning using privileged information. In: Proceedings of the 2013 International Joint Conference on Neural Networks (IJCNN), pp. 1–8. Dallas, Texas, USA, Aug 2013
7. Fullerton, A.S., Xu, J.: The proportional odds with partial proportionality constraints model for ordinal response variables. Soc. Sci. Res. **41**(1), 182–198 (2012)
8. Gutiérrez, P.A., Pérez-Ortiz, M., Sánchez-Monedero, J., Fernández-Navarro, F., Hervás-Martínez, C.: Ordinal regression methods: survey and experimental study. IEEE Trans. Knowl. Data Eng. **28**(1), 127–146 (2016)
9. Herbrich, R., Graepel, T., Obermayer, K.: Support vector learning for ordinal regression. In: Proceedings of the 9th International Conference on Artificial Neural Networks (ICANN), pp. 97–102. Edinburgh, UK, Sep 1999
10. McCullagh, P.: Regression models for ordinal data. J. Roy. Stat. Soc. B (Methodol.) **42**(2), 109–127 (1980)
11. Nguyen, B., Morell, C., De Baets, B.: Distance metric learning for ordinal classification based on triplet constraints. Knowl.- Based Syst. **142**, 17–28 (2018)
12. Nguyen, Q., Valizadegan, H., Hauskrecht, M.: Learning classification with auxiliary probabilistic information. In: Proceedings of the 11th International Conference on Data Mining (ICDM), pp. 477–486. IEEE, Vancouver, BC (2011)

13. Nguyen, Q., Valizadegan, H., Hauskrecht, M.: Learning classification models with soft-label information. J. Am. Med. Inform. Assoc. **21**(3), 501–508 (2014)
14. Nguyen, Q., Valizadegan, H., Seybert, A., Hauskrecht, M.: Sample-efficient learning with auxiliary class-label information. In: AMIA Annual Symposium Proceedings, vol. 2011, p. 1004. American Medical Informatics Association (2011)
15. PASCAL: (Pattern Analysis, Statistical Modelling and Computational Learning) machine learning benchmarks repository. http://mldata.org/ (2011)
16. Tang, M., Pérez-Fernández, R., De Baets, B.: Fusing absolute and relative information for augmenting the method of nearest neighbors for ordinal classification. Inform. Fusion **56**, 128–140 (2020)
17. Waegeman, W., De Baets, B., Boullart, L.: Learning to rank: a ROC-based graph-theoretic approach. Pattern Recogn. Lett. **29**(1), 1–9 (2008)
18. Weinberger, K.Q., Saul, L.K.: Distance metric learning for large margin nearest neighbor classification. J. Mach. Learn. Res. **10**(2), 207–244 (2009)
19. Xue, Y., Hauskrecht, M.: Efficient learning of classification models from soft-label information by binning and ranking. In: Proceedings of the 30th International Florida Artificial Intelligence Research Society Conference, pp. 164–169. Marco Island, Florida, May 2017

Towards Deterministic Diverse Subset Sampling

J. Schreurs$^{(\boxtimes)}$, M. Fanuel, and J. A. K. Suykens

Department of Electrical Engineering (ESAT), STADIUS Center for Dynamical
Systems, Signal Processing and Data Analytics, KU Leuven,
Kasteelpark Arenberg 10, 3001 Leuven, Belgium
{joachim.schreurs,michael.fanuel,johan.suykens}@kuleuven.be

Abstract. Determinantal point processes (DPPs) are well known models for diverse subset selection problems, including recommendation tasks, document summarization and image search. In this paper, we discuss a greedy deterministic adaptation of DPPs. Deterministic algorithms are interesting for many applications, as they provide interpretability to the user by having no failure probability and always returning the same results. First, the ability of the method to yield low-rank approximations of kernel matrices is evaluated by comparing the accuracy of the Nyström approximation on multiple datasets. Afterwards, we demonstrate the usefulness of the model on an image search task.

Keywords: Determinantal point processes · Landmark sampling · Diversity

1 Introduction

Selecting a diverse subset is an interesting problem for many applications. Examples are document or video summarization [3,10,13,14], image search tasks [11], pose estimation [13] and many others. Diverse sampling algorithms have also shown their benefits to calculate a low-rank matrix approximations using the Nyström method [21]. This method is a popular tool for scaling up kernel methods, where the quality of the approximation relies on selecting a representative subset of landmark points or Nyström centers.

Notations. In this work, we will use uppercase letters for matrices and calligraphic letters for sets, while bold letter denote random variables. The notation $(\cdot)^\dagger$ denotes the Moore-Penrose pseudo inverse of a matrix. We also define the partial order of positive definite (resp. semidefinite) matrices by $A \succ B$ (resp. $A \succeq B$) if and only if $A - B$ is positive definite (resp. semidefinite). Furthermore, we denote by K, the Gram matrix $[k(x_i, x_j)]_{i,j=1}^n$ obtained from a positive semidefinite kernel such as the Gaussian kernel $k(x, y) = \exp(-\|x - y\|_2^2/(2\sigma^2))$.

Nyström Approximation. The Nyström method takes a positive semidefinite matrix $K \in \mathbb{R}^{n \times n}$ as input, selects from it a small subset $\mathcal{C}$ of columns, and constructs the approximation $\hat{K} = K_{\mathcal{C}} K_{\mathcal{C}\mathcal{C}}^\dagger K_{\mathcal{C}}^\top$, where $K_{\mathcal{C}} = KC$ and $K_{\mathcal{C}\mathcal{C}} = C^\top KC$

B. Bogaerts et al. (Eds.): BNAIC 2019/BENELEARN 2019, CCIS 1196, pp. 137–151, 2020.
https://doi.org/10.1007/978-3-030-65154-1_8

are submatrices of the kernel matrix and $C \in \mathbb{R}^{n \times |\mathcal{C}|}$ is a sampling matrix obtained by selecting the columns of the identity matrix indexed by $\mathcal{C}$. The matrix $\hat{K}$ is used in the place of K, so to decrease the training runtime and memory requirements. Using a dependent or *diverse* sampling algorithm for the Nyström approximation has shown to give better performance than independent sampling methods in [8, 15].

Determinantal Point Processes and Kernel Methods. Determinantal point processes (DPPs) [12] are well known models for diverse subset selection problems. A point process on a ground set $[n] = \{1, 2, ..., n\}$ is a probability measure over point patterns, which are finite subsets of $[n]$. It is common to define a DPP thanks to its marginal kernel, that is a positive symmetric semidefinite matrix satisfying $P \preceq \mathbb{I}$. Let $\mathcal{Y}$ denote a random subset, drawn according to the DPP with marginal kernel P. Then, the probability that $\mathcal{C}$ is a subset of the random $\mathcal{Y}$ is defined by

$$\Pr(\mathcal{C} \subseteq \mathcal{Y}) = \det(P_{\mathcal{C}\mathcal{C}}). \tag{1}$$

Notice that all principal submatrices of a positive semidefinite matrix are positive semidefinite. From (1), it follows that:

$$\Pr(i \in \mathcal{Y}) = P_{ii}$$
$$\Pr(i, j \in \mathcal{Y}) = P_{ii}P_{jj} - P_{ij}P_{ji}$$
$$= \Pr(i \in \mathcal{Y})\Pr(j \in \mathcal{Y}) - P_{ij}^2.$$

The diagonal elements of the kernel matrix give the marginal probability of inclusion for individual elements, whereas the off-diagonal elements determine the "repulsion" between pairs of elements. Thus, for large values of P_{ij}, or a high similarity, points are unlikely to appear together. In some applications, it can be more convenient to define DPPs thanks to L-ensembles, which can be related to marginal kernels by the formula $L = P(\mathbb{I} - P)^{-1}$ when $P \prec \mathbb{I}$ (more details in [2]). They allow to define the probability of sampling a random subset $\mathcal{Y}$ that is equal to $\mathcal{C}$:

$$\Pr(\mathcal{Y} = \mathcal{C}) = \frac{\det(L_{\mathcal{C}\mathcal{C}})}{\det(\mathbb{I} + L)}. \tag{2}$$

In contrast to (1), the only requirement on L is that it has to be positive semidefinite. Notice that the normalization in (2) can be derived classically by considering the property relating the coefficients of the characteristic polynomial of a matrix to the sum of the determinant of its principal submatrices of the same size. In this paper, the L-ensemble is chosen to be a kernel matrix K.

Exact sampling of a DPP is done in two phases [12]. Let V be the matrix whose columns are the eigenvectors of K. In the first phase, a subset of eigenvectors of the kernel matrix K is selected at random, where the probability of selecting each eigenvector depends on its associated eigenvalue in a specific way given in Algorithm 1. In the second phase, a sample $\mathcal{Y}$ is produced based on the selected vectors. At each iteration of the second loop, the cardinality of $\mathcal{Y}$ increases by one and the number of columns of V is reduced by one. A k-DPP [11] is a DPP conditioned on a fixed cardinality $|\mathcal{Y}| = k$. Note that $e_i \in \mathbb{R}^n$ is the i-th vector of the canonical basis.

---

input: L-ensemble $L \succeq 0$
initialization: $\mathcal{J} = \emptyset$ and $\mathcal{Y} = \emptyset$
Calculate the eigenvector/value pairs $\{(v_i, \lambda_i)\}_{i=1}^n$ of L.
for: $i = 1, \ldots, n$ **do**
    $\mathcal{J} \leftarrow \mathcal{J} \cup \{i\}$ with prob. $\frac{\lambda_i}{\lambda_i + 1}$
end for
$V \leftarrow \{v_i\}_{i \in \mathcal{J}}$ a set of columns
while: $|V| > 0$ **do**
    Draw an index i according to the distribution $p_i = \frac{1}{|V|} \sum_{v \in V} (v^{\mathrm{T}} e_i)^2$.
    $\mathcal{Y} \leftarrow \mathcal{Y} \cup i$
    $V \leftarrow V_\perp$, an orthonormal basis for the subspace of V orthogonal to e_i.
end while
return $\mathcal{Y}$.

Algorithm 1: Exact DPP sampling algorithm associated to the L-ensemble L [12]. Notice that $P = L(L + \mathbb{I})^{-1}$, so that the eigenvector/value pairs of P are exactly $\{(v_i, \frac{\lambda_i}{\lambda_i + 1})\}_{i=1}^n$.

Deterministic algorithms are interesting for many applications, as they provide interpretability to the user by having no chance of failure and always returning the same results. The usefulness of deterministic algorithms has already been recognized by Papailiopoulos et al. [18] and McCurdy [17], who provide deterministic algorithms based on the (ridge) leverage scores. These statistical leverage scores correspond to correlations between the singular vectors of a matrix and the canonical basis [1,7]. The recently introduced Deterministic Adaptive Sampling (DAS) algorithm [8] provides a deterministically obtained *diverse* subset. The method shows superior performance compared to randomized counterparts in terms of approximation error for the Nyström approximation when the eigenvalues of the kernel matrix have a fast decay. A similar observation was made for the deterministic algorithms of Papailiopoulos et al. [18] and McCurdy [17].

This paper discusses a deterministic adaption of k-DPP, where we have the following empirical observations:

1. The method is deterministic, hence there is no failure probability and the method always produces the same output.
2. Only the k eigenvectors with the largest eigenvalues are needed, which results in a speedup when $k \ll n$.
3. We observed that the method samples a more diverse subset than the original k-DPP on multiple datasets.
4. There is no need to tune a regularization parameter, which is the case for the DAS algorithm.

5. The method shows superior accuracy in terms of the max norm of the Nyström approximation on multiple datasets compared to the standard k-DPP, along with better accuracy of the kernel approximation for the operator norm when there is fast decay of the eigenvalues.

In Sect. 2, we introduce the method. Secondly, we make a connection with the DAS algorithm, namely the deterministic k-DPP corresponds to the DAS algorithm with an adapted projector kernel matrix. In Sect. 3, we evaluate the method on different datasets. Finally, a small real-life illustration is shown in Sect. 4.

2   Deterministic Adaptation of k-DPP

We discuss a deterministic adaptation of k-DPP, by selecting iteratively landmarks with the highest probability. As it is described in Algorithm 2, we can successively maximize the probability over a nested sequence of sets $C_0 \subseteq C_1 \subseteq \cdots \subseteq C_k$ starting with $C_0 = \emptyset$ by adding one landmark at each iteration. The proposed method is an adaptation of the improved k-DPP sampling algorithm given by Tremblay et al. [20]. Our proposed method start from a projective marginal kernel $P = VV^\top$, with $V = [v_1, ..., v_k] \in \mathbb{R}^{n \times k}$ the sampled eigenvectors of the kernel matrix. Instead of sampling the eigenvectors [12], the k eigenvectors with the largest eigenvalue are chosen. Secondly, at each iteration the point with the highest probability $p(i) = P_{ii} - P_{Ci}^\top P_{CC}^\dagger P_{Ci}$ is chosen, where C corresponds to the selected subset so far. Besides the interpretation of DPPs in relation to diversity, the aforementioned probability gives a second insight in diversity. Namely we have $p(i) = \|v_i - \pi_{V_C} v_i\|_2^2$, where $v_i \in \mathbb{R}^{n \times 1}$ is the i-th column of V and π_{V_C} is the projector on $V_C = \text{span}\{v_s | s \in C\}$. The chosen landmark corresponds to the point that is the most distant to the space of the previously sampled points.

---

input: Kernel matrix K, sample size k.
initialization: $C \leftarrow \emptyset$
Calculate the first k eigenvectors $V \in \mathbb{R}^{n \times k}$ from K.
$P = VV^\top$
Define $p_0 \in \mathbb{R}^N : \forall i, \quad p_0(i) = \|V^\top e_i\|^2$
$p \leftarrow p_0$
for: $i = 1, \ldots, k$ **do**
    Select c_i with highest probability $p(i)$
    $C \leftarrow C \cup \{c_i\}$
    Update $p : \forall j \quad p(j) = p_0(j) - P_{Cj}^\top P_{CC}^\dagger P_{Cj}$
end for
return C.

Algorithm 2: Deterministic adaptation of the k-DPP sampling algorithm.

2.1 Connections with DAS

Algorithmically, the proposed method corresponds to the DAS algorithm [8] (see, Algorithm 3 in appendix) with a different projector kernel matrix. More precisely, DAS uses a smoothed projector kernel matrix $P_{n\gamma}(K) = K(K+n\gamma I)^{-1}$ with $K \succ 0$. The ridge leverage scores $l_i(\gamma) = \sum_{j=1}^{n} \frac{\lambda_j}{\lambda_j+n\gamma} V_{ij}^2$ can be found on the diagonal $P_{n\gamma}(K)$. Let $V \in \mathbb{R}^{n \times n}$ be the matrix of eigenvectors of the kernel matrix K. On the contrary, in this paper, the proposed method has a *sharp projector kernel matrix* with the rank-k leverage scores $l_i = \sum_{j=1}^{k} V_{ij}^2$ on the diagonal. This has the added benefit that there is no regularization parameter to tune. The DAS algorithm is given in the Appendix.

Algorithm 2 is a greedy reduced basis method as defined in [6]. Other greedy maximum volume approaches are found in [4,5,9]. These greedy methods are used for finding an estimate of the most likely configuration (MAP), which is known to be NP-hard [9]. In practice, we see that the method performs quite well and gives a consistently larger $\det(K_{CC})$, which is considered a measure for diversity, compared to the randomized counterpart (see Sect. 3). This number is calculated as follows $\log(\det(K_{CC})) = \sum_{i=1}^{k} \log(\lambda_i)$, with $\{\lambda_i\}_{i=1}^{k}$ the singular values of K_{CC}. A small illustration is given in Fig. 1, where the deterministic algorithm gives a more diverse subset.

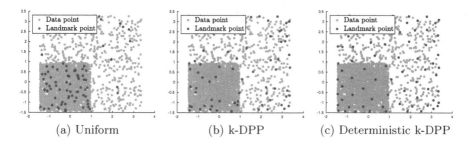

(a) Uniform   (b) k-DPP   (c) Deterministic k-DPP

Fig. 1. Illustration of sampling methods on an artificial dataset. Uniform sampling does not promote diversity and selects almost all points in the bulk of the data. Sampling a k-DPP overcomes this limitation, however landmarks can be close to each other. The latter is solved by using the deterministic adaptation of k-DPP, which gives a more diverse subset.

3 Numerical Results

We evaluate the performance of the deterministic variant of the k-DPP with a Gaussian kernel on the `Boston housing`, `Stock`, `Abalone` and `Bank 8FM` datasets, which have 506, 950, 4177 and 8192 datapoints respectively. Those public datasets[1] have been used for benchmarking k-DPPs in [15]. The implementation of the algorithms is done with MatlabR2018b.

Throughout the experiments, we use a fixed bandwidth $\sigma = 2$ for **Boston housing**, **Stock** and **Abalone** dataset and $\sigma = 5$ for the **Bank 8FM** dataset after standardizing the data. The following algorithms are used to sample k landmarks: Uniform sampling, k-DPP[2] [11], DAS [8] and the proposed method. DAS is executed for multiple regularization parameters $\gamma \in \{10^0, 10^{-1}, \ldots, 10^{-6}\}$ where the sample with the best performing γ is selected to approximate the kernel matrix. The total experiment is repeated 10 times. The quality of the landmarks $\mathcal{C}$ is evaluated by the relative operator or max norm $\|K - \hat{K}\|_{\{\infty,2\}}/\|K\|_{\{\infty,2\}}$ with $\hat{K} = K_\mathcal{C}(K_{\mathcal{C}\mathcal{C}} + \epsilon\mathbb{I})^{-1}K_\mathcal{C}^\top$ with $\varepsilon = 10^{-12}$ for numerical stability. The max norm and the operator norm of a matrix A are given respectively by $\|A\|_\infty = \max_{i,j}|A_{ij}|$ and $\|A\|_2 = \max_{\|x\|_2=1}\|Ax\|_2$. The diversity is measured by $\log(\det(K_{\mathcal{C}\mathcal{C}}))$, where a larger log determinant means more diversity. The results for the **Stock** dataset are visible in Fig. 2. The results for the rest of the datasets or shown in Figs. 8, 9, 10 and 11 in the Appendix. The computer used for these simulations has 8 processors 3.40 GHz and 15.5 GB of RAM.

As previously mentioned, the greedy method returns a more diverse subset. Figure 8 shows the $\log(\det(K_{\mathcal{C}\mathcal{C}}))$ where the proposed method shows similar performance as DAS, while improving on both the randomized methods. The same is visible for the relative max norm of the Nyström approximation error. DAS and the deterministic variant of the k-DPP perform well on the **Boston housing** and **Stock** dataset, which show a fast decay in the spectrum of K (see Fig. 7). If the decay of the eigenvalues is not fast enough, the randomized k-DPP, shows better performance. The same observation was made for the deterministic (ridge) leverage score sampling algorithms [17,18] as well as DAS [8].

4   Illustration

We demonstrate the use of the proposed method on a image summarization task. The first experiment is done on the **Stanford Dogs** dataset[3], which contains images of 120 breeds of dogs from around the world. This dataset has been built using images and annotation from ImageNet for the task of fine-grained image categorization. The training features are SIFT descriptors [16] (given by the dataset), which are used to make a histogram intersection kernel. We take a subset of 50 images each of the classes *border collie*, *chihuahua* and *golden retriever*. The total training set is visualized in Fig. 5 in the Appendix. Figure 3 displays the results of the method for $k = 4$. One can observe that the images are very dissimilar and dogs out of each breed are represented. This is confirmed by the projection of the landmarks on the 2 first principal components of the

---

[1] https://www.cs.toronto.edu/~delve/data/datasets.html, https://www.openml.org/d/223.

[2] We used the Matlab code available at https://www.alexkulesza.com/.

[3] http://vision.stanford.edu/aditya86/ImageNetDogs/main.html.

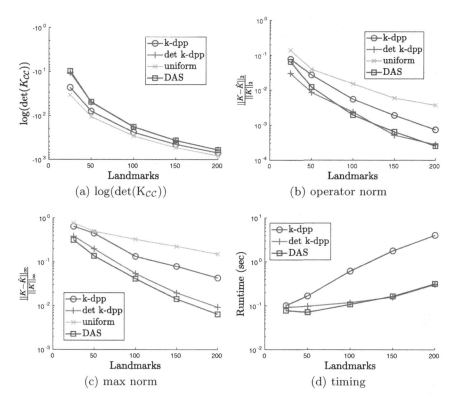

Fig. 2. The $\log(\det(K_{CC}))$, relative operator norm and relative max norm of the Nyström approximation error and timings as a function of the number of landmarks on the Stock dataset. The results are plotted on a logarithmic scale, averaged over 10 trials. The larger $\log(\det(K_{CC}))$, the more diverse the subset.

kernel principal component analysis (KPCA) [19], where the landmark points lie in the outer regions of the space.

We repeat the above procedure on the `Kimia99` dataset[4]. The dataset has 9 classes consisting of 11 images each. It contains shapes silhouettes for the classes: rabbits, quadrupeds, men, airplanes, fish, hands, rays, tools, and a miscellaneous class. The total training set is visible in Fig. 6 in the Appendix. First, we resize the images to size 100×100. Afterwards, we apply a Gaussian kernel with bandwidth $\sigma = 100$ after standardizing the data. The results of the proposed method with $k = 9$ are visible in Fig. 4. The method samples landmarks out every class, making it a desirable image summarization. This is supported by the projection of the landmarks on the 2 first principal components of the KPCA, where a landmark points is chosen out of every small cluster.

---

[4] https://vision.lems.brown.edu/content/available-software-and-databases# Datasets-Shape.

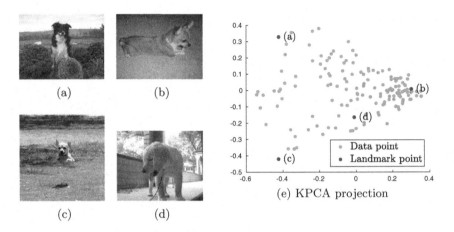

Fig. 3. Illustration of the proposed method with $k = 4$ on the **Stanford Dogs** dataset. The selected landmark points are visualized on the left, the projection on the 2 first principal components of the KPCA on the right.

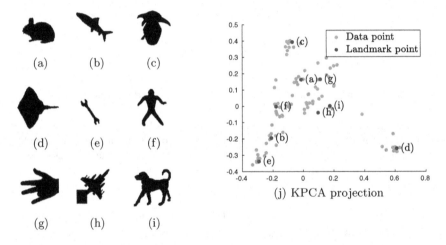

Fig. 4. Illustration of the proposed method with $k = 9$ on the **Kimia99** dataset. The selected landmark points are visualized on the left, the projection on the 2 first principal components of the KPCA on the right.

5   Conclusion

We discussed a greedy deterministic adaptation of k-DPPs. Algorithmically, the method corresponds to the DAS algorithm with a different projector kernel matrix. The proposed method is evaluated by comparing the accuracy of the Nyström approximation on multiple datasets. Experiments show the proposed method is able to give a more diverse subset, along with better performance for the relative max norm. When there is a fast decay of the eigenvalues, the deterministic method is more accurate than randomized counterparts. To conclude, we demonstrate the usefulness of the model on an image search task.

Acknowledgements. EU: The research leading to these results has received funding from the European Research Council under the European Union's Horizon 2020 research and innovation program/ERC Advanced Grant E-DUALITY (787960). This paper reflects only the authors' views and the Union is not liable for any use that may be made of the contained information. Research Council KUL: Optimization frameworks for deep kernel machines C14/18/068 Flemish Government: FWO: projects: GOA4917N (Deep Restricted Kernel Machines: Methods and Foundations), PhD/Postdoc grant Impulsfonds AI: VR 2019 2203 DOC.0318/1QUATER Kenniscentrum Data en Maatschappij Ford KU Leuven Research Alliance Project KUL0076 (Stability analysis and performance improvement of deep reinforcement learning algorithms).

A   Additional Algorithms

input: Matrix $K \succ 0$, sample size k and $\gamma > 0$.
initialization: $\mathcal{C}_0 = \emptyset$ and $m = 1$.
$P \leftarrow K(K + n\gamma \mathbb{I})^{-1}$
while: $m \leq k$ **do**
  $s_m \in \arg\max diag \left(P - P_{\mathcal{C}_m} P_{\mathcal{C}_m \mathcal{C}_m}^{-1} P_{\mathcal{C}_m}^{\top} \right)$.
  $\mathcal{C}_m \leftarrow \mathcal{C}_{m-1} \cup \{s_m\}$ and $m \leftarrow m + 1$.
end while
return $\mathcal{C}_m$.

Algorithm 3: DAS algorithm [8].

B   Additional Figures

Fig. 5. The training data of the Stanford Dogs dataset.

Fig. 6. The training data of the `Kimia99` dataset.

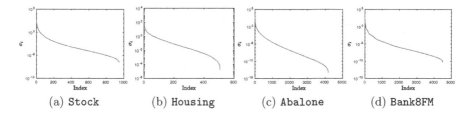

(a) Stock   (b) Housing   (c) Abalone   (d) Bank8FM

Fig. 7. Singular value spectrum of the datasets on a logarithmic scale. For a given index, the value of the eigenvalues for the `Stock` and `Housing` dataset are smaller than `Abalone` and `Bank8FM`.

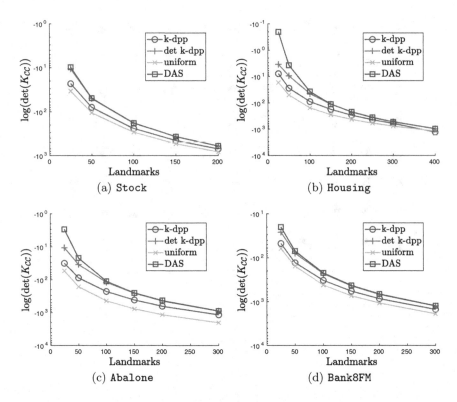

Fig. 8. $\log(\det(K_{CC}))$ in function of the number of landmarks. The error is plotted on a logarithmic scale, averaged over 10 trials. The larger the $\log(\det(K_{CC}))$, the more diverse the subset

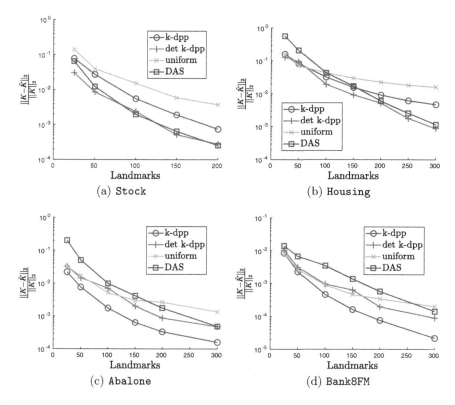

Fig. 9. Relative operator norm of the Nyström approximation error as a function of the number of landmarks. The error is plotted on a logarithmic scale, averaged over 10 trials.

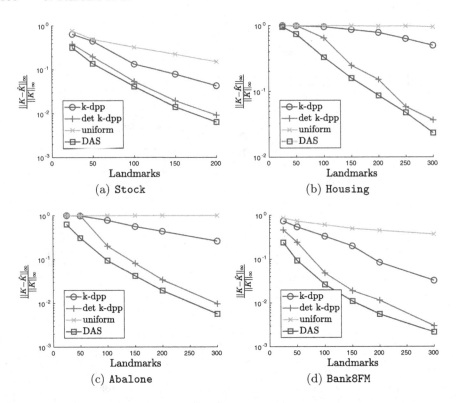

Fig. 10. Relative max norm of the approximation as a function of the number of landmarks. The error is plotted on a logarithmic scale, averaged over 10 trials.

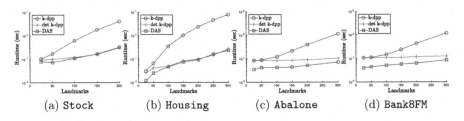

Fig. 11. Timings for the computations of Fig. 9 as a function of the number of landmarks. The timings are plotted on a logarithmic scale, averaged over 10 trials.

References

1. Alaoui, A., Mahoney, M.W.: Fast randomized kernel ridge regression with statistical guarantees. In: Advances in Neural Information Processing Systems, pp. 775–783 (2015)
2. Borodin, A.: Determinantal point processes. arXiv preprint arXiv:0911.1153 (2009)
3. Carbonell, J.G., Goldstein, J.: The use of MMR, diversity-based reranking for reordering documents and producing summaries. SIGIR **98**, 335–336 (1998)

4. Chen, L., Zhang, G., Zhou, E.: Fast greedy map inference for determinantal point process to improve recommendation diversity. In: Advances in Neural Information Processing Systems, pp. 5622–5633 (2018)
5. Çivril, A., Magdon-Ismail, M.: On selecting a maximum volume sub-matrix of a matrix and related problems. Theor. Comput. Sci. **410**(47–49), 4801–4811 (2009)
6. DeVore, R., Petrova, G., Wojtaszczyk, P.: Greedy algorithms for reduced bases in banach spaces. Construct. Approx. **37**(3), 455–466 (2013)
7. Drineas, P., Magdon-Ismail, M., Mahoney, M.W., Woodruff, D.P.: Fast approximation of matrix coherence and statistical leverage. J. Mach. Learn. Res. **13**, 3475–3506 (2012)
8. Fanuel, M., Schreurs, J., Suykens, J.A.K.: Nyström landmark sampling and regularized Christoffel functions. arXiv preprint arXiv:1905.12346 (2019)
9. Gillenwater, J., Kulesza, A., Taskar, B.: Near-optimal map inference for determinantal point processes. In: Advances in Neural Information Processing Systems, pp. 2735–2743 (2012)
10. Gong, B., Chao, W.L., Grauman, K., Sha, F.: Diverse sequential subset selection for supervised video summarization. In: Advances in Neural Information Processing Systems, pp. 2069–2077 (2014)
11. Kulesza, A., Taskar, B.: k-DPPs: fixed-size determinantal point processes. In: Proceedings of the 28th International Conference on Machine Learning, pp. 1193–1200 (2011)
12. Kulesza, A., Taskar, B.: Determinantal point processes for machine learning. Found. Trends Mach. Learn. **5**(2–3), 123–286 (2012)
13. Kulesza, A., Taskar, B.: Structured determinantal point processes. In: Advances in Neural Information Processing Systems, pp. 1171–1179 (2010)
14. Kulesza, A., Taskar, B.: Learning determinantal point processes. arXiv preprint arXiv:1202.3738 (2012)
15. Li, C., Jegelka, S., Sra, S.: Fast DPP sampling for Nyström with application to kernel methods. In: Proceedings of the 33rd International Conference on International Conference on Machine Learning, pp. 2061–2070 (2016)
16. Lowe, D.G., et al.: Object recognition from local scale-invariant features. In: ICCV, vol. 99, pp. 1150–1157 (1999)
17. McCurdy, S.: Ridge regression and provable deterministic ridge leverage score sampling. In: Advances in Neural Information Processing Systems, vol. 31, pp. 2468–2477 (2018)
18. Papailiopoulos, D., Kyrillidis, A., Boutsidis, C.: Provable deterministic leverage score sampling. In: Proceedings of the 20th ACM SIGKDD International Conference on Knowledge Discovery and Data Mining, pp. 997–1006 (2014)
19. Schölkopf, B., Smola, A., Müller, K.-R.: Kernel principal component analysis. In: Gerstner, W., Germond, A., Hasler, M., Nicoud, J.-D. (eds.) ICANN 1997. LNCS, vol. 1327, pp. 583–588. Springer, Heidelberg (1997). https://doi.org/10.1007/BFb0020217
20. Tremblay, N., Barthelme, S., Amblard, P.O.: Optimized algorithms to sample determinantal point processes. arXiv preprint arXiv:1802.08471 (2018)
21. Williams, C.K., Seeger, M.: Using the Nyström method to speed up kernel machines. In: Advances in Neural Information Processing Systems, pp. 682–688 (2001)

Extended Bayesian Personalized Ranking Based on Consumption Behavior

Alireza Gharahighehi[1,2(✉)] and Celine Vens[1,2]

[1] Itec, imec research group at KU Leuven, Kortrijk, Belgium
`alireza.gharahighehi@kuleuven.be, celine.vens@kuleuven.be`
[2] Faculty of Medicine, KU Leuven, Campus KULAK, Kortrijk, Belgium

Abstract. Bayesian Personalized Ranking (BPR) is a well-known recommendation framework that learns to rank items based on one-class implicit feedback. In some domains such as video and music streaming and news aggregator websites, users' implicit feedback is not limited to one-class feedback as there are other types of feedback such as watching, listening and reading time which are continuous. This feedback reflects the consumption behavior of users. In this research we show that using this kind of implicit feedback on the top of one-class feedback, recommender systems are able to learn user preferences more precisely. We propose an extended form of BPR by including user consumption behavior to recommend news topics. The result shows that using the extended form of BPR with consumption information improves the performance based on four evaluation measures. The result also verifies that by considering a more granular feedback the extended BPR has better predictions.

Keywords: News recommendation · Implicit feedback · Bayesian personalized ranking

1 Introduction

Users of digital services continuously interact with websites, mobile applications and personal wearables. These interactions produce a huge amount of information about user experiences and preferences, which can be used by service providers to personalize their services and improve user experience and satisfaction. Service providers use recommender systems to provide more relevant services to the users.

To recommend relevant items, recommender systems use feedback from users. This feedback could be explicit like rating, liking and disliking, or implicit like clicking on an item url, adding an item to basket, and purchasing. In most cases only implicit feedback is available because users are not willing to provide their explicit feedback on the items. A recommender system should rely on the existing implicit feedback from a user in the website when the explicit feedback from that user is missing.

B. Bogaerts et al. (Eds.): BNAIC 2019/BENELEARN 2019, CCIS 1196, pp. 152–164, 2020.
https://doi.org/10.1007/978-3-030-65154-1_9

One of the well-known recommender system approaches based on implicit feedback is Bayesian Personalized Ranking (BPR) [1]. BPR is a learning to rank recommendation framework that uses user specific pair-wise preferences as training data. The main assumption in this framework is that users prefer observed items to unobserved items. Therefore BPR only considers binary implicit feedback.

In some digital services user's implicit feedback is more granular than binary feedback. For instance, consumption behaviour such as watching, listening and reading time are real valued types of implicit feedback, reflecting user preferences to a video, a music track or an article. The higher the consumption level the more interest a user has in the consumed item. In this article we propose to use this type of information as an additional preference indication about observed items.

Item recommendation is not always the only goal of recommender systems. Some times digital services want to recommend higher-level products such as news topics, music play-lists and video channels. Consumption behavior of users on items can be used to recommend higher-level contents to the user [2].

In this article we propose an extended version of BPR using consumption behavior of users on news articles to recommend news topics in a news aggregator website. In this extended version, additional pair-wise comparisons between observed items are added to the training data.

In the following, related studies about implicit feedback and BPR are reviewed (Sect. 2). In Sect. 3 we explain the BPR framework and show how the user consumption behavior is considered in a proposed extended version of BPR. In Sect. 4 the used datasets are described and the experimental settings in designing and evaluating the model are discussed. Then the effect of considering consumption behavior in BPR is assessed in Sect. 5 and discussed in Sect. 6. Finally, we conclude in Sect. 7 and outline future work in Sect. 8.

2   Related Work

News is highly dynamic, i.e., the set of items changes rapidly. Most of the news recommenders focus on this characteristic and try to consider recent trends and user short-term interests in their models [3,4]. Recommending news articles is not the only goal of news aggregator websites though. In some of these websites users can subscribe to some topics and later on they will receive personalized news articles based on these subscribed topics. The set of news topics is not highly dynamic and therefore complex matrix completion methods can be used to recommend news topics to the users.

BPR [1] is a matrix completion framework to design recommender systems based on user-specific relative preferences between observed and unobserved items. For instance, if a user observed the webpage of item 'A' and did not observe the webpage of item 'B', then BPR assumes the user prefers item 'A' to item 'B'. It uses bootstrap sampling to draw samples from user-specific pairwise comparisons to train the model. Each sample consists of an observed item and an unobserved item for a user. Stochastic gradient descent is used to update model parameters for each training sample. This sampling-updating strategy is continued until convergence.

There are some BPR extensions which add confidence values to pair-wise comparisons. Wang et al. [5] added a confidence score to the BPR framework in a Tweet recommender system. This score measures users' confidence on their pairwise preferences based on received time of tweets. Pan et al. [6] argued that users' implicit feedback is not always homogeneous. They used confidence weights for two levels of implicit feedback. For instance, in a web-shop purchasing an item implies higher confidence than viewing an item web-page.

Instead of considering only observed and unobserved status for items, multiple levels of preferences can be used based on implicit feedback. Loni et al. [7] proposed and extended BPR for multi-channel implicit feedback. They define multiple channels of feedback consisting of three main categories: Observed positive feedback, unobserved and observed negative feedback. Then they defined a sampler to draw user-specific pair-wise samples from these multiple channels. Lerche and Jannach [8] claimed that in real-world applications there are different types of user actions such as view, add to basket and purchase. They proposed a framework to draw pair-wise comparisons from these different levels of observed items. Yu et al. [9] divided the unobserved items into different levels and proposed a multiple pair-wise ranking involving multiple items instead of simple pairwise ranking.

BPR is not the only framework for implicit feedback. In this study we compared the performance of proposed method with two strong baselines: Weighted Regularized Matrix Factorization (WRMF) [10,11] and Weighted Approximate-Rank Pairwise (WARP) [12]. WRMF is a collaborative filtering approach for implicit feedback datasets. WARP was initially proposed for image annotation but was later on used as a recommender system.

There are some news datasets that provide users' logs on news articles. Zhang [13] conducted a user study to collect explicit and implicit feedback and users' actions on news articles. Gulla et al. [14] prepared the Adressa dataset that contains user interactions with Norwegian news articles. The dataset includes user implicit feedback on news articles, contextual data and articles metadata. The study of these news datasets is considered as a future work.

3   Topic Recommender

BPR learns its model parameters based on pairwise comparisons between observed and unobserved items. For instance if we use matrix factorization in the BPR framework, the model parameters are users and items low-rank matrices. Matrix factorization is a matrix completion method that predicts the interaction matrix between users and items by constructing user and item representations, i.e, U and I, in a same dimensional space. The predicted interaction matrix $\hat{X}$ can be calculated by $\hat{X} = U \times I^T$ where U is a $n \times k$ matrix, I is a $m \times k$ matrix, n is the number of users, m is the number of items and k is the number of latent factors.

Algorithm 1 shows the intuition behind BPR as presented in [1]. Each training sample (u,i,j) in D_s contains a user id u, a positive item i and a negative item j for that user. BPR assumes that this user prefers item i to item j. The Stochastic Gradient Descent (SGD) approach is used to update parameters based on the training tuples. $\hat{x}_{uij}$ is predicted by the model and is an arbitrary value which captures the relationship between user u, item i and item j. If we use matrix factorization in the BPR framework, $\hat{x}_{uij}$ is calculated by:

$$\hat{x}_{uij} = U_u \cdot I_i^T - U_u \cdot I_j^T \tag{1}$$

in which U and I are user and item low-rank matrices, i.e., the parameters to be learned. In Algorithm 1, α is the learning rate and λ is the regularization term.

Algorithm 1 Bayesian Personalized Ranking (BPR), as presented in [1]

Input: Training samples $(u, i, j) \in D_S$
Output: Learned parameters Θ (user and item low-rank matrices)
Initialize parameters Θ
Repeat
$\quad$ draw a sample (u,i,j) from D_S
$\quad \Theta \leftarrow \Theta + \alpha(\dfrac{e^{-\hat{x}_{uij}}}{1 + e^{-\hat{x}_{uij}}} \cdot \dfrac{\partial}{\partial \Theta}\hat{x}_{uij} + \lambda_\Theta \Theta)$
Until *Convergence*;

In this research we propose an extended form of BPR (EBPR) to recommend news topics to the users based on their consumption behaviour on news articles. EBPR has exactly the same steps like BPR (Algorithm 1 and Eq. 1) but with an extended training set. The new instances in this extended training set are generated based on users' consumption behaviour. In news aggregator websites this consumption behaviour consists of clicking, dwelling time in an article webpage, scroll-depth and timestamp of the event. Dwelling time and scroll-depth show to what extent the user consumed the item and timestamp reflects the recency of this consumption. A similar setting exists in other domains such as music and video streaming websites. In these websites listening or watching time can be used as the consumption behaviour.

To calculate the consumption level we use the following function:

$$c_{ui} = \begin{cases} dt_{ui} \cdot sd_{ui} \cdot t_{ui} & \text{if user } u \text{ clicked article } i \\ 0 & \text{otherwise} \end{cases} \tag{2}$$

where c_{ui} is the consumption level of user u for item i and dt_{ui}, sd_{ui} and t_{ui} are the values for dwelling time, scroll-depth and timestamp respectively for user u and item i. We compute these values in a relative way, by defining a maximal consumption level and then comparing the actual consumption level to the maximal consumption level. In our case maximal consumption level is 100 percent scroll-depth, length-based minimum reading time and current timestamp. Length-based minimum reading time depends on the number of words in the article. By specifying a standard reading time for a word, minimum reading time for an article can be estimated by its length. Reading time can depend on the reading speed of a specific user, but there is no need to normalize reading time based on the user speed, as in BPR the positive and negative items of a training sample are both from the same user. A user has a maximal consumption level for an article if she spends enough time, scrolls the whole page and does this very recently. High dwelling time without scrolling and high scroll-depth with very low dwelling time reflect very low consumption.

To reflect the level of consumption in the feedback weight, i.e, to calculate dt_{ui}, sd_{ui} and t_{ui}, a decay function should be used. This function decays the feedback weight based on the difference between the maximal and actual consumption level. Various decay functions such as linear (Eq. 3) and exponential (Eq. 4) can be used to reflect the difference between actual and maximal values.

$$f(x_{ui}) = max(0, 1 - \beta \cdot \frac{X_{max} - x_{ui}}{X_{max}}) \tag{3}$$

$$f(x_{ui}) = e^{-\beta \cdot (\frac{X_{max} - x_{ui}}{X_{max}})} \tag{4}$$

where $f(x_{ui})$ is the feedback weight (dt_{ui}, sd_{ui} or t_{ui}), X_{max} is the maximal consumption level of variable x for user u, x_{ui} is the actual value and β is a positive constant that controls the level of decay. For instance, if we consider the feedback weight of *one* for the maximal consumption of an article, and if the user scrolled only 20%, then the feedback weight should be decayed based on 80% difference. In this example $sd_{ui} = max(0, 1 - 0.8\beta)$. For timestamp, X_{max} is the timestamp of the most recent interaction of user u in the system.

The aim of topic recommendation is to recommend news topics to the users based on their consumption behavior on news articles. By aggregating user consumption behavior at the article level, user preferences on news topics can be inferred. A normalization function should be applied on aggregated values. On the one hand, there are some niche topics which have few related articles and on the other hand there are general topics with a large number of related articles. In Eq. (5), n_k is the number of articles related to the topic k. A straightforward form of f_{norm} is dividing $\sum_{i \in k} c_{ui}$ by n_k.

$$r_{uk} = f_{norm}(\sum_{i \in k} c_{ui}, n_k) \tag{5}$$

Based on the calculated user preferences on news topics (r_{uk}), additional training instances can be used and the BPR framework can be applied to recommend news topics to the users. The EBPR has the same steps like BPR (Algorithm 1 and Eq. 1) but the training data contains additional instances. Table 1 shows the intuition behind the additional training instances in EBPR. In this table (Table 1b and 1c) "+" and "-" show the preference for the topic in the corresponding column versus the topic in the corresponding row and "?" means that the user does not have any preferences between these two topics. For instance (see Table 1a), *user#150* has observed *Topic*1, *Topic*2 and *Topic*5 and has not observed *Topic*3, *Topic*4 and *Topic*6. Based on the original BPR (see Table 1b) $user\#150$ prefers $Topic1$ to $Topic3$, $Topic4$ and $Topic6$ because $Topic1$ is observed and the other ones are not observed. This user does not have any preferences between $Topic2$ and $Topic1$ and also $Topic5$ and $Topic1$ because all of them are observed. Using the calculated user preferences (r_{uk} in EBPR) (see Table 1c), we assume that this user prefers $Topic2$ to $Topic1$ and $Topic1$ to $Topic5$.

Table 1. Illustration of additional training data using calculated user preferences

(a) Calculated preferences of User#150 for the observed topics

	Topic1	Topic2	Topic3	Topic4	Topic5	Topic6
User#150	121.52	247.5	Missing	Missing	50.8	Missing

(b) The training data for User#150 based on observed/ not observed topics in BPR

	Topic1	Topic2	Topic3	Topic4	Topic5	Topic6
Topic1		?	−	−	?	−

(c) The additional training data for User#150 based on calculated preferences in EBPR

	Topic1	Topic2	Topic3	Topic4	Topic5	Topic6
Topic1		+	−	−	−	−

Adding pairs from observed items, one should consider sparsity. Most of the datasets are sparse which means the number of observed items is much smaller than unobserved items. Therefore drawing negative samples uniformly from the whole dataset leads to almost the original BPR. To exploit this newly added information one can specify the portion of training data which should be drawn from positive items.

4   Dataset and Experiments Set-Up

We use three datasets to evaluate the presented approach.[1] Two datasets from a Flemish news content aggregator website (Roularta Media Group) and a music dataset. In one of the news datasets (News-1), for each user-article interaction, information about consumption behavior such as binary click, dwelling time, scroll-depth and timestamp is given over a period of 60 days. The other news dataset (News-2) is given over a period of 200 days and it contains the same information, apart from scroll depth, which is lacking. Based on this given information, we want to recommend news topics to the users. Relations between articles and news topics are also given. To evaluate our method on a publicly available dataset, we use Lastfm-2k (http://www.lastfm.com)[2] to recommend artists that users listened to their songs and tags that users assigned to the artists. To the best of our knowledge, there is no publicly available dataset consisting of consumption behavior such as listening or watching time. Therefore

---

[1] The source code is available on https://itec.kuleuven-kulak.be/supportingmaterial.

[2] The dataset can be obtained from https://grouplens.org/datasets/hetrec-2011/.

we consider the number of interactions per user-item pair as an indication of consumption behavior, i.e., more interactions reflect more consumption in the music dataset. Table 2 describes these datasets. In this table items are higher-level contents and #items refers to number of artists and tags in the music dataset and number of news topics in the news datasets.

There are some hyperparameters that are tuned. We use matrix factorization [15] as the model in the BPR framework. Therefore hyparameters are the number of latent factors, regularization term λ_Θ, learning rate α, and the parameter λ of the decay function. To tune these hyperparameters one interaction per user is kept out of the training set and used as the validation set. The hyperparameters are tuned based on the model performance ($nDCG@10$) in the validation set using a grid search.

To evaluate the presented framework we use four evaluation measures; AUC, $nDCG@k$, $precision@k$ and $recall@k$. AUC measures the ability of model to correctly compare pairs of items. To calculate AUC, similar to sampling in BPR training, we draw a positive and a negative item from the test set and check whether the trained model is able to prefer an observed item to an unobserved one, i.e., predicts higher score for the observed item. Normalized Discounted Cumulative Gain (nDCG) is a learning-to-rank evaluation method which considers higher penalties for irrelevant items in the recommendation list if they are ranked at the very beginning of the ranking list. $precison@k$ and $recall@k$ are standard information retrieval performance measures. $precision@k$ is the proportion of recommendation list that is relevant and $recall@k$ is the proportion of relevant items in the recommendation list. For the test set we keep two random positive topics of each user for evaluation. Training set and test set are completely disjoint.

Table 2. Dataset descriptions

	Lastfm (Artist)	Lastfm (Tag)	News-1	News-2
# Users	1,892	1,860	476	1,337
# Items	12,523	9,749	996	2,638
# Interactions	71,080	35,178	6,163	56,432
Sparsity	0.003	0.002	0.013	0.016

We compare four models: extended BPR (EBPR), BPR, WARP and WRMF.[3] Unlike BPR, in EBPR we draw pair-wise comparisons not only from observed and unobserved items, but also from observed items. For the news datasets we consider two versions of EBPR. In the first version (EBPR_I) the consumption level for each user-item consists of number of interactions and in the second version (EBPR_II) consumption level contains more granular information such as number of interactions, reading time, scroll-depth and timestamp.

---

[3] Mrec library (https://mendeley.github.io/mrec/) is used for WARP and WRMF.

As explained before, for the music datasets, only (EBPR_I) can be used. WARP and WRMF are two popular recommendation algorithms based on implicit feedback (see Sect. 2). Popularity-based and random-based recommendations are also included as baselines.

5 Results

Table 3, Table 4, Table 5 and Table 6 show the results for the artist, tag and news topic recommendations respectively. As is shown in Table 3 (artist recommendation), the extended form of BPR (EBPR) performs better than original BPR in all performance measures. The additional information about consumption behavior of users assists BPR to have better predictions compared to the original version. EBPR is also the best performing approach compared to the other baselines based on three performance measures. WARP performs better compared to WRMF and has better AUC compared to BPR.

According to Table 4 (tag recommendation) EBPR again performs better than original BPR based on all four performance measures. It is the best performing method in three measures compared to the other baselines. This means that by using the additional information provided by consumption levels the model has better predictions of user preferences. WARP has a better performance compared to BPR.

Table 5 shows the results for news topic recommendation (News-1). The first version of EBPR (EBPR_I) which uses the number of interactions as consumption levels has better performance in all four measures compared to original BPR. Using more granular information, i.e., dwelling time and scroll-depth, EBPR_II has a better performance compared to EBPR_I and original BPR. It is also the best performing approach compared to the other baselines in this dataset. WARP performs better than original BPR in three measures.

In the News-2 dataset, again extended versions have better performance compared to BPR except for AUC. According to Table 6, EBPR_II is the best performing approach in three performance measures compared to other baselines. It again shows that the performance of EBPR is higher when the more granular information such as dwelling time are added to the model. In this dataset WARP and WRMF perform better compared to BPR.

Concerning model complexity, the extended version of BPR does not affect the complexity of the original algorithm. EBPR provides more possible training examples compared to BPR. In BPR the number of possible training examples for each user is $\mathcal{O}(|I_u^+|.|I_u^-|)$ and for the extended versions it is $\mathcal{O}(|I_u^+|.|I|)$ where $|I_u^+|$ is the number of interactions of user u, $|I_u^-|$ is the number of items that user u did not interact with and $|I|$ is the number of items. The original version tends to converge faster compared to the extended BPR. As depicted in Fig. 1, BPR converges faster compared to EBPR but results in a lower performance.

Table 3. Results of artist recommendation

	AUC	nDCG@10	Precision@10	Recall@10
BPR	0.863	0.246	0.045	0.192
EBPR_I	0.878	**0.305**	**0.053**	**0.201**
WARP	**0.887**	0.224	0.038	0.188
WRMF	0.870	0.177	0.031	0.153
Random	0.517	0.002	0.001	0.002
Popular	0.838	0.032	0.011	0.057

Table 4. Results of tag recommendation

	AUC	nDCG@10	Precision@10	Recall@10
BPR	0.831	0.297	0.052	0.263
EBPR_I	0.839	**0.321**	**0.055**	**0.277**
WARP	**0.841**	0.302	**0.055**	0.275
WRMF	0.838	0.286	0.053	0.266
Random	0.496	0.011	0.002	0.011
Popular	0.828	0.123	0.036	0.180

6   Discussion

The main assumption of this research is, the more a user consumes an item and the more recent this consumption, the higher preference she has on that item. Considering this, one can add pairwise comparisons between observed items to the BPR training data. The original BPR assumes that users only prefer observed items to unobserved items. The EBPR uses the additional information based on consumption levels to form additional pairwise preferences for each user.

The consumption behavior of users can be easily logged in many contexts such as video and music streaming websites and also text content aggregators. This feedback reflects more detailed information than binary observed/unobserved implicit feedback. This additional information aids the model to learn more based on relations between observed items and performs better than the original BPR. Based on the results presented in Sect. 5, EBPR has relatively better performance not only compared to BPR but also compared to two strong baselines, WARP and WRMF.

Table 5. Results of news recommendation (News1)

	AUC	nDCG@10	Precision@10	Recall@10
BPR	0.911	0.589	0.116	0.538
EBPR_I	0.914	0.690	0.132	0.609
EBPR_II	**0.933**	**0.754**	**0.138**	**0.678**
WARP	0.921	0.686	0.114	0.572
WRMF	0.897	0.567	0.103	0.517
Random	0.514	0.019	0.003	0.015
Popular	0.853	0.045	0.043	0.208

Table 6. Results of news recommendation (News2)

	AUC	nDCG@10	Precision@10	Recall@10
BPR	0.916	0.415	0.078	0.397
EBPR_I	0.912	0.426	0.095	0.475
EBPR_II	0.913	**0.527**	**0.097**	**0.481**
WARP	0.926	0.489	0.088	0.442
WRMF	**0.934**	0.471	0.085	0.427
Random	0.503	0.004	0.001	0.004
Popular	0.904	0.033	0.045	0.224

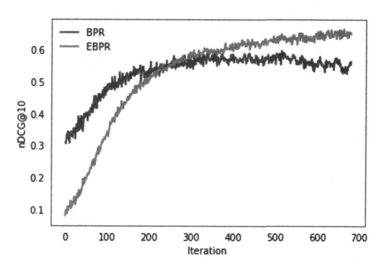

Fig. 1. The convergence behavior of BPR and EBPR

The results in Table 5 and Table 6 confirmed that by using more granular consumption levels the model is able to predict the user preferences more precisely. This also confirms the general assumption of our study that users prefer items that are more consumed. Our extension does not affect the complexity of BPR as it has a similar algorithmic approach but with some additional training data. The idea of this study can be used to recommend higher level contents such as news topics, tags, music playlists, artists and video channels based on consumption behavior of users on items such as articles, music tracks and videos.

7   Conclusion

Implicit feedback from users is not usually limited to binary feedback (observed/unobserved). There is more granular information about user feedback which should be considered in recommendation systems. Consumption behavior such as number of clicks, reading time and scroll-depth are usually available in news websites and service providers should use this information to have a better prediction model for user preferences. This additional information shows to what extent a user consumes and consequently prefers a particular item. More clicks, longer reading time and higher scroll-depth rate in item web-page show higher preferences.

In this study we proposed an extended form of Bayesian Personalized Ranking (BPR) to include user specific pair-wise comparisons between observed items in the training set using consumption behavior of users. Using a publicly available datatset (lastfm) and two real-world implementation news datasets, in almost all cases the extended form of BPR outperforms the original BPR based on four measures: AUC, nDCG precision and recall. Our method also has a better performance compared to two other strong baselines, WARP and WRMF.

8   Future Directions

For future work we outline the following directions:

- We will use the calculated user preference for each topic to adapt the gradient magnitude in the stochastic gradient descent update. In BPR the gradient magnitude does not depend on user feedback.
- To address the cold-start problem one can use existing side information such as user profiles, interactions between users [16] and item features [17]. We will assess the effect of other types of side information such as item co-occurrence graphs and hierarchical structure in BPR.
- We will assess the effect of adding other consumption information such as the device that the user used and the medium through which the interaction was triggered.
- The proposed method can be applied on other news datasets such as YOW [13] and Adressa [14].

Acknowledgments. This work was executed within the imec.icon project NewsButler, a research project bringing together academic researchers (KU Leuven, VUB) and industry partners (Roularta Media Group, Bothrs, ML6). The NewsButler project was co-financed by imec and received project support from Flanders Innovation & Entrepreneurship (project nr. HBC.2017.0628).

References

1. Rendle, S., Freudenthaler, C., Gantner, Z., Schmidt-Thieme, L.: BPR: Bayesian personalized ranking from implicit feedback. arXiv preprint arXiv:1205.2618 (2012)
2. Liang, H., Xu, Y., Tjondronegoro, D., Christen, P.: Time-aware topic recommendation based on micro-blogs. In: Proceedings of the 21st ACM International Conference on Information and Knowledge Management, pp. 1657–1661 (2012)
3. Jugovac, M., Jannach, D., Karimi, M.: StreamingRec: a framework for benchmarking stream-based news recommenders. In: Proceedings of the 12th ACM Conference on Recommender Systems, pp. 269–273 (2018)
4. Karimi, M., Jannach, D., Jugovac, M.: News recommender systems-survey and roads ahead. Inf. Process. Manage. **54**(6), 1203–1227 (2018)
5. Wang, S., Zhou, X., Wang, Z., Zhang, M.: Please spread: recommending tweets for retweeting with implicit feedback. In: Proceedings of the 2012 Workshop on Data-Driven User Behavioral Modelling and Mining from Social Media, pp. 19–22 (2012)
6. Pan, W., Zhong, H., Congfu, X., Ming, Z.: Adaptive Bayesian personalized ranking for heterogeneous implicit feedbacks. Knowl.-Based Syst. **73**, 173–180 (2015)
7. Loni, B., Pagano, R., Larson, M., Hanjalic, A.: Bayesian personalized ranking with multi-channel user feedback. In: Proceedings of the 10th ACM Conference on Recommender Systems, pp. 361–364 (2016)
8. Lerche , L., Jannach, D.: Using graded implicit feedback for Bayesian personalized ranking. In: Proceedings of the 8th ACM Conference on Recommender Systems, pp. 353–356 (2014)
9. Yu, R., et al.: Multiple pairwise ranking with implicit feedback. In: Proceedings of the 27th ACM International Conference on Information and Knowledge Management, pp. 1727–1730 (2018)
10. Hu, Y., Koren, Y., Volinsky, C.: Collaborative filtering for implicit feedback datasets. In: 2008 Eighth IEEE International Conference on Data Mining, pp. 263–272. IEEE (2008)
11. Pan, R., et al.: One-class collaborative filtering. In: 2008 Eighth IEEE International Conference on Data Mining, pp. 502–511. IEEE (2008)
12. Weston, J., Bengio, S., Usunier, N.: WSABIE: scaling up to large vocabulary image annotation. In: Twenty-Second International Joint Conference on Artificial Intelligence (2011)
13. Zhang, Y.: Bayesian graphical models for adaptive filtering. In: SIGIR Forum, vol. 39, pp. 57 (2005)
14. Gulla, J.A., Zhang, L., Liu, P., Özgöbek, Ö., Su, X.: The adressa dataset for news recommendation. In: Proceedings of the International Conference on Web Intelligence, pp. 1042–1048 (2017)
15. Sarwar, B., Karypis, G., Konstan, J., Riedl, J.: Incremental singular value decomposition algorithms for highly scalable recommender systems. In: Fifth International Conference on Computer and Information Science, vol. 1, pp. 27–8. CiteSeer (2002)

16. Krohn-Grimberghe, A., Drumond, L., Freudenthaler, C., Schmidt-Thieme, L.: Multi-relational matrix factorization using Bayesian personalized ranking for social network data. In: Proceedings of the Fifth ACM International Conference on Web Search and Data Mining, pp. 173–182 (2012)
17. Gantner, Z., Drumond, L., Freudenthaler, C., Rendle, S., Schmidt-Thieme, L.: Learning attribute-to-feature mappings for cold-start recommendations. In: 2010 IEEE International Conference on Data Mining, pp. 176–185. IEEE (2010)

SubSect—An Interactive Itemset Visualization

Joey De Pauw[1(✉)] (ID), Sandy Moens[1] (ID), and Bart Goethals[1,2] (ID)

[1] University of Antwerp, Antwerp, Belgium
{joey.depauw,sandy.moens,bart.goethals}@uantwerpen.be
[2] Monash University, Melbourne, Australia

Abstract. Itemsets and association rules are among the most simple and intuitive patterns that are being used to explore transaction datasets. However, they lack meaning without both context and domain knowledge. Typically a user has to sift through hundreds of these patterns before finding an interesting one, losing sight of the forest for the trees. We propose a novel itemset and association rule visualization that makes it possible to inspect, assess, and compare patterns at a glance. In a case study we demonstrate its ability to facilitate a user in deriving and presenting valuable insights from a real-world dataset, which can not only save time and effort, but also reduce errors introduced by misconceptions.

Keywords: Visualization · Pattern mining · Itemsets · Association rules

1 Introduction

Pattern mining is a commonly used technique in data exploration and data analysis [1]. In contrast to actively querying the data, pattern mining has the advantage of letting the data tell you what it looks like. Essentially, patterns such as itemsets and association rules provide an efficient way to represent local structures in the data. Most importantly, they have a summarizing property which facilitates the end user in interpreting and understanding a dataset.

Unfortunately, pattern mining alone does not suffice: typically a large number of patterns exists, even for relatively small datasets, making the process of discovering truly *interesting* patterns very tedious and strenuous for the practitioner. A transaction dataset with 20 different items for example, contains 2^{20} (more than 1 million) candidate itemsets. This is known as the pattern explosion problem. To make matters worse, *interestingness* is a subjective measure that can only be approximated by objective metrics or features [15].

In previous work this problem has been tackled for instance by sorting and filtering patterns based on different metrics [6] or by trying to minimize the number of reported patterns to the most descriptive subset [3,17]. Another approach is to represent patterns in informative visualizations and rely on the end

© Springer Nature Switzerland AG 2020
B. Bogaerts et al. (Eds.): BNAIC 2019/BENELEARN 2019, CCIS 1196, pp. 165–181, 2020.
https://doi.org/10.1007/978-3-030-65154-1_10

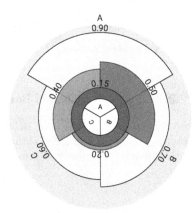

Fig. 1. Example of the visualization for an arbitrary itemset $\{A, B, C\}$.

user to find what is interesting in their respective domain [4,5,9–13,16]. Our contribution is situated in the latter context.

We propose a visualization for itemsets based on the double decker plot from Hofmann et al. [11]. It exploits the monotonicity property which states that itemsets have a lower or equal support compared to the support of their subsets. An example of our visualization for the arbitrary itemset $\{A, B, C\}$ can be seen in Fig. 1. To demonstrate the power of this visualization, we integrated the JavaScript-based implementation in the data mining and visualization tool SNIPER (formerly known as MIME [8]) and performed a case study.

This paper is organized as follows. In Sect. 2 we provide the required background in pattern mining and visualization. Section 3 describes the visualization itself with a theoretical analysis. Section 4 includes a case study where the efficiency of the visualization is verified in practice. Related work is discussed in Sect. 5 and finally we conclude our work in Sect. 6.

2   Background

2.1   Pattern Mining

Pattern mining is the process of discovering statistically relevant patterns in large datasets [7]. We focus on the mining of itemsets in a transaction database with items $\mathcal{I}$. A transaction database is a collection of subsets of $\mathcal{I}$. The *support* of an itemset is defined as the number of transactions that contain the itemset:

$$Supp(X) = |\{t \in \mathcal{D} \mid X \subseteq t\}|$$

with $\mathcal{D}$ the transaction database, t a transaction and X an itemset. Frequency is defined as the relative support:

$$Freq(X) = \frac{suppX}{supp\emptyset}$$

For every itemset a range of association rules can be derived by splitting it in two parts: an antecedent X and a consequent Y. An association rule is denoted as $X \rightarrow Y$, where both X and Y are itemsets and $X \cap Y = \varnothing$. We define the *support* and *confidence* of an association rule as follows:

$$Supp(X \rightarrow Y) = Supp(X \cup Y) \qquad Conf(X \rightarrow Y) = \frac{Supp(X \rightarrow Y)}{Supp(X)}$$

Confidence is the conditional probability of a transaction containing itemset Y when X is already present. Additionally, we define *lift* as the ratio of observed support to that expected if X and Y were independent:

$$Lift(X \rightarrow Y) = \frac{Supp(X \rightarrow Y)}{Supp(X) \times Supp(Y)}$$

A frequent pattern mining algorithm such as Apriori or Eclat can be used to find all itemsets with a support higher than some user defined minimum support threshold [7]. This threshold is imposed to limit the search to patterns that are statistically relevant. With a minimum support of 1 the algorithm would simply calculate every pattern that occurs in the dataset, whereas with the minimum support threshold set to the number of transactions, the algorithm would only report patterns that occur in every transaction.

It is clear that the choice of minimum support has a big impact on the results of the pattern mining step and unfortunately there is usually no simple way to find the "right" value for this parameter. Therefore, choosing this parameter is typically an iterative and highly interactive process: a higher value may be too restrictive whereas a lower value can result in more uninteresting patterns that clutter the output. Defining the interestingness of a pattern usually requires domain knowledge which classic algorithms cannot take into account [15].

Alternatively, one can work bottom-up; whereas the first approach first mines an abundance of patterns followed by a filtering step, the bottom-up approach relies on patterns being built from the ground up by a domain expert. Hybrid solutions on the other hand try to include the user in each stage of the mining process, where for example the set of candidate itemsets can be reduced or expanded at each iteration of Apriori before the algorithm continues [18] or a framework is provided for the user to edit, combine or augment various patterns from different techniques [8].

Visualization plays a key role in any of these approaches. In the purely algorithmic approach, visualizations are mostly used in the filtering step, where a concise but informative visualization of itemsets is preferred over a plain list of itemsets. The bottom-up and hybrid approaches on the other hand use visualizations both for finding interesting combinations of items as well as for inspecting the resulting itemsets.

2.2   Visualization

The main advantage of using visualizations over textual representations is that they allow for better perceptual processing [14]. There are two phases in the theory of information processing: *perceptual processing* (seeing) and *cognitive processing* (understanding). Perceptual processes are automatic, very fast, and mostly executed in parallel, while cognitive processes operate under conscious control of attention and are relatively slow, effortful, and sequential [14].

As an example of the power of perceptual processing one can imagine finding all itemsets that contain a specific item. In a textual notation you are limited to reading all the itemsets and remembering which ones contained the desired item. In a notation where all items are given a distinctive color it becomes possible to identify all occurrences of the item with a glance over the visualization. Likewise, many other visual variables can be used to encode information into the visualization, facilitating the user in extracting the desired information. This effect becomes even more powerful when considering comparisons between entities.

Visual variables can be categorized in *planar variables* (horizontal and vertical position) and *retinal variables* (shape, size, color, brightness, orientation and texture). A visualization can make use of these eight primitives in the visual alphabet to encode information, however it is not always desirable to use all of them. Leaving some degree of freedom is helpful when combining visualizations or annotating them.

In our contribution *shape, size, color, brightness* and *orientation* are all used as part of the visualization to varying degrees. *Texture* and *position* are left free for annotation and for integration in other tools. The relative position of instances of the visualization is not defined, which allows it to be used in more complex layouts or potentially even in other visualization techniques.

3   SubSect

We present an efficient visualization for displaying itemsets and association rules. Its main goal is to show the most relevant information about an itemset (or a collection of itemsets) and allow for a domain expert to quickly interpret whether or not it is interesting. For this purpose, intuitiveness is very important.

Keeping the theory of visualization in mind, we define the following goals for our visualization in the context of pattern mining:

G1. The most important properties of a pattern should be semantically clear, i.e. readable on the visualization in an unambiguous way, ideally through perceptual processing.
G2. Users should be able to combine their domain knowledge with properties of itemsets to discover the most interesting patterns.
G3. A user should be able to easily compare two patterns.
G4. Ideally, more insights can be derived from the visualization or from the combination of two or more instances. For example, when the itemsets $\{A, B, C\}$ and $\{A\}$ are visible, information about the itemset $\{A, B\}$ may also be derived. Or from the set $\{A, B, C\}$ information about the rule $\{A, B\} \rightarrow \{C\}$ can be inferred.

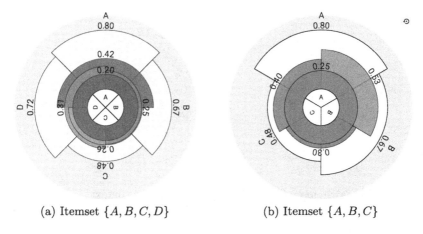

(a) Itemset $\{A, B, C, D\}$ (b) Itemset $\{A, B, C\}$

Fig. 2. The visualization for an arbitrary itemset (a) and for one of its subsets (b).

3.1 Basic Usage

To explain how our visualization works, we first consider the example in Fig. 2a. Every item in the itemset is represented in the center. The arcs around the center items show three levels of itemsets that can be formed from these items: 1) the itemset containing all k items, 2) all k-1 itemsets and 3) all singleton itemsets. For example, the blue full circle includes all four items A, B, C and D, and has a frequency of 0.2 as indicated by the label and its radius. The other segments represent subsets, like for example the cyan arc which spans items A, B and C. In correspondence with the higher frequency of this itemset (0.25), its arc also has a proportionally larger radius.

In order to reduce the overlap between arcs, we have chosen to let them span between the centers of the outer two items, rather than to have them cover 100% of the edge items. This trade-off reduces image clutter and therefore improves the scalability of our visualization, at the cost of a steeper learning curve: new users would expect the full items to be covered, but with some practice we believe that the meaning of the segments does become intuitive. The fact that same-cardinality itemsets always have the same *shape* (arc width) helps in this respect. Additionally, by hovering over the arc, label or frequency value of an itemset, all three of them are emphasized, clarifying which visual elements belong together.

Furthermore, in every image only the most interesting and informative subsets are rendered: for a k-itemset these are the k-1-itemsets and the singleton itemsets. In the left example for an itemset of 4 items, we display the 4-itemset, the 3-itemsets and the singleton itemsets. Together this combination of subsets provides the most useful information: the singleton itemsets give a global context and the k-1-itemsets place the k-itemset in a local context. Note that the arcs for singleton itemsets are always shaded in white while the others are given a unique color per itemset. This makes it easier to link multiple instances of the visualization that have items in common (G3).

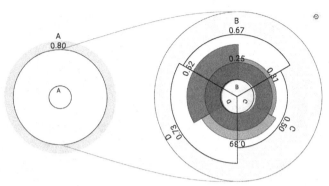

(a) Itemset $\{B, C, D\}$ in $\{A\}$-conditional context.

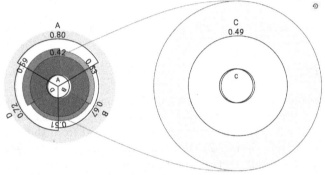

(b) Itemset $\{C\}$ in $\{A, B, D\}$-conditional context.

Fig. 3. A more advanced example with the α-conditional view.

Finally, the visualization is equipped with three interactions to maximize usability: *dive deeper*, *α-conditional view* and *reset*. Animations like hover highlighting indicate the presence of these interactions and gradual transition animations ease the transition between "states" of the visualization, making the effect of the interactions more clear. Clicking on the cyan arc for example will *dive into* its respective itemset $\{A, B, C\}$. An animation shows that item D is removed from the center and the cyan arc becomes a full circle. Three new subsets are now visible. The result is shown in Fig. 2b. Naturally this action can be repeated from the new view to *dive deeper* or the user can choose to go back to the top level with the *reset* button that just became available.

In the next Sect. 3.2 we demonstrate a more advanced use case with the *α-conditional view*. Section 3.3 discusses the visualization from a theoretical point of view and analyzes its strengths and weaknesses.

3.2   Alpha-Conditional View

For any given set of items α we define the α-conditional database as the set of transactions that contain all items in α. It provides for a natural way of thinking about association rules. When we say *"80% of the people who buy diapers also buy beer"*, we essentially say that the itemset $\{beer\}$ has a relative support of 80% in the $\{diapers\}$-conditional database or equivalently that the association rule $\{diapers\} \rightarrow \{beer\}$ has a confidence of 80%.

Similar to the interaction for selecting an itemset to dig deeper, it is also possible to click a single item (in the center or on the outer edges) and add it to the α set or the *"scope"* as can be seen in Fig. 3a. In this α-conditional view, the scope is always visible on the smaller visualization to the left. On the right-hand side, we see the remaining items and itemsets, but now with their frequencies relative to the scope.

By moving more items to the scope, it becomes clear that the scope set is rendered as another instance of the itemset visualization, i.e. with its respective subsets (see Fig. 3b). This makes for a very interesting synergy, since not only the scope is visible, but also the context of what the α-conditional transactions look like. Again one can reset the visualization back to its original state with the reset button. Additionally it is possible to click items or itemsets in the left visualization to expand the scope by moving items back to the right-hand side.

3.3   Theoretical Analysis

Itemsets. It is obvious that the frequency of the entire itemset, its k-1 subsets and its constituent items can be seen trivially through the labels and radii (G1). By the monotonicity property however, we can also derive some information about all the itemsets in between (G4). This is especially useful when the bounds are close together, since this provides a tighter estimate. In Fig. 2a for example, we can derive that the frequency of $\{D, C\}$ must lie between 0.31 and 0.48 through the itemsets $\{A, D, C\}$ and $\{C\}$, since they are the most frequent superset and the least frequent subset respectively.

Association Rules. The relation between an itemset and its subsets also implies an association rule. If two arcs are close together in terms of their radii, we know the association rule they imply will also have a high confidence (G4). Recall that the formula for confidence is $\frac{Supp(\{A,B,C,D\})}{Supp(\{A,B,C\})}$ for the rule $\{A, B, C\} \rightarrow \{D\}$. In Figure 2a we find the frequencies are 0.20 and 0.25, leading to a confidence of 0.8, which is also intuitively *"guessable"* from the difference in radius; i.e. without calculating the value, it is also simple to estimate it quite accurately from the visualization.

More importantly, we also introduced the α-conditional view to facilitate representing association rules. The frequency in an α-conditional database is equivalent to the confidence of the rule with α as antecedent and the itemset as consequent. Hence we can form any association rule from the itemset by simply moving the antecedent items to the scope and *"browsing"* to the desired itemset on the right-hand side. For example, Fig. 3a shows, among others, the association rule $\{A\} \rightarrow \{B, C, D\}$ with a confidence of 0.25. We can also see that $\{A\}$ has a support of 0.8 and, from this, derive that $Supp(\{A\} \rightarrow \{B, C, D\}) = 0.8 \times 0.25 = 0.2$.

Scalability. Despite our efforts to reduce overlap and clutter, it remains infeasible to render itemsets that consist of a large amount of items. For a k-itemset, $2k + 1$ arcs are rendered, which is already more favourable than an exponential amount. However, an inherent issue arises from the k-1-subsets, whose arcs need to span $k - 1$ items, resulting in an inhibiting amount of overlap for large k.

Specific solutions can be considered to facilitate the visualization of large patterns, such as combining items together in a preprocessing step or manually selecting the subsets to be rendered, for example by grouping some frequent and less interesting items in the center (see Sect. 6.1). In our experience however, large patterns are often of limited use: they are more complex to understand and typically either have a low frequency or contain many correlated/very frequent attributes that do not contribute to the pattern.

When the interactions between many attributes need to be studied, we opt for a collection of smaller patterns (that have attributes in common) over a single large pattern. Our visualization is better suited for this methodology of combining multiple instances together (G3).

After this brief analysis we find that our visualization already succeeds at goals G1, G3 and G4. Goal G2 relates to domain specific knowledge being integrated, which we did not discuss yet. The case study in the next section (Sect. 4) illustrates this concept.

4   Case Study

To demonstrate the effectiveness of our visualization we performed a case study with a real-world dataset. The main goal of this case study is to show how the visualization assists the end user in finding interesting patterns based on their domain knowledge (i.e. goal G2 from Sect. 3). A customer churn dataset[1] that describes information about customers of a telecom company who left within the last month was used. We chose this dataset because it is easy to understand without requiring any specific expertise or background (as for example would be the case for a financial or political dataset), yet it shows some interesting patterns. In total it contains 21 columns, describing information about $7,043$ customers. Table 1 documents a subset of these attributes, i.e. the ones that appear in our examples. For every attribute we give the icon, name, description and its possible values.

First, the dataset was loaded in the research tool SNIPER[2], a web-based tool for pattern mining with a main focus on facilitating data exploration [8]. In the

---

[1] https://www.kaggle.com/blastchar/telco-customer-churn/.
[2] https://bitbucket.org/sandymoens/sniper/.

Table 1. Icon, name, description and possible values for each attribute in the dataset.

Attribute	Values
☗ Churn	Yes No
Indicates whether the customer left within the last month. This is the intended target variable of the dataset.	
👫 Partner	Yes No
Yes if the customer has a partner, *No* otherwise.	
👤 Dependents	Yes No
Whether or not the customer has dependents. In most cases dependents are children, students or elderly people.	
📋 Contract	Month-to-month One year Two year
The contract term of the customer.	
🌐 Internet Service	Fiber optic DSL No
Which type of internet service the customer opted for, or *No* if none.	
📞 Phone Service	Yes No
Indicates if the contract includes phone service.	
⚥ Gender	Male Female
The gender of the customer.	
🔒 Online Security	Yes No No internet service
Whether the customer has the online security service or *No internet service* if N/A.	
💬 Tech Support	Yes No No internet service
Similar to security, this attribute indicates if the contract includes tech support.	

setup phase, we defined icons for each attribute and decided on a discretization strategy to handle numeric variables. Five equal-width buckets were used. Given the context, this choice provided a good resolution with adequate support for the individual items. After preprocessing, the resulting transaction dataset consisted of 60 unique items and 7,043 transactions.

Then the dataset was explored with the functionality provided by SNIPER. This includes, but is not limited to, classical itemset mining (like Eclat [7]), rule mining, sorting and filtering of patterns and manually building patterns by combining them or forming them based on the insight brought by various metrics. We mainly used the latter technique to create patterns based on the conditional support and lift metrics.

The following sections each provide examples of patterns that were found in the data and how our visualization was used to find and interpret them. A live version for each example can be found on https://joeydp.github.io/SubSect/.

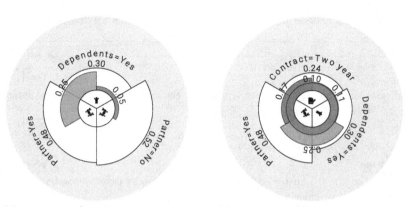

(a) Itemset {Partner=Yes, Part-
ner=No, Dependents=Yes}

(b) Itemset {Partner=Yes, Con-
tract=Two year, Dependents=Yes}

Fig. 4. Example of two interesting itemsets rendered in our visualization.

4.1 Example - Lift

The most straightforward use of our visualization is illustrated in this example. Suppose we are interested in the attribute *partner* and would like to investigate if there is a relation with the attribute *dependents*. Figure 4a gives a concise and intuitive representation of the information needed to compare these two attributes. Note that we assume the meaning and possible values of these attributes are known beforehand, i.e. we know these are two boolean attributes with mutually exclusive values.

First we can see that a little over half of the customers with a partner, also have dependents ($\frac{0.25}{0.48}$). For customers without a partner this ratio is only one in ten ($\frac{0.05}{0.52}$). Without context this information is quite meaningless, so we compare it to the expected distribution of dependents, which is 30% for the overall dataset. Now it becomes clear that customers with a partner have a higher chance of also having dependents and inversely the chance is lower for customers without a partner. In the other direction we can see that five out of six customers with dependents also have a partner and the remaining one out of six do not.

Both of the previously described patterns show association rules with relatively high confidence. The advantage of using this visualization is that the lift, i.e. the support divided by the expected support if the variables were independent, can also easily be derived. This is a good example to illustrate why the single itemsets and k-1 itemsets were selected to be visualized: the local and global context synergize to allow the end user to derive new information.

The second example in Fig. 4b shows the presence of an association rule with a very high confidence, albeit with a relatively low support. That is 11% of the customers have a *two year contract* and *dependents*, and 10% have those two and a *partner*, leading to he association rule {*Dependents, Two year contract*} → {*Partner*} with a confidence of around 91%, which is intuitive given the domain

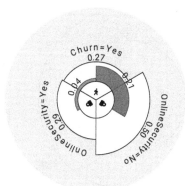

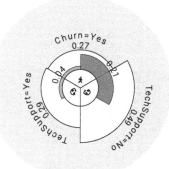

(a) Itemset {Churn=Yes, Online Se-
curity=No, Online Security=Yes}

(b) Itemset {Churn=Yes, Tech Sup-
port=No, Tech Support=Yes}

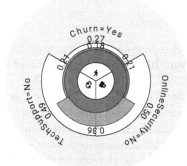

(c) Itemset {Churn=Yes, Tech Sup-
port=No, Online Security=No}

Fig. 5. Example to show how multiple instances of the visualization can be used
together and how the dependence between attributes can be derived.

knowledge behind these attributes. It seems logical that customers with a long
term contract and with dependents would be more likely to have a partner.

Perhaps more interesting is the fact that there is relatively little overlap
between the two association rules that constitute the previous one. We find
that only 25 in 30 people with dependents also have a partner ($\approx$83%) and
that only $\frac{17}{24} \approx 71\%$ of the people with a two year contract have a partner.
In other words, both variables "contribute" to the association rule in the sense
that without either one, the confidence would drop. This information can all be
derived intuitively from our visualization.

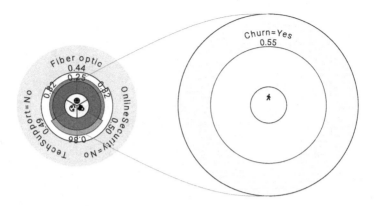

Fig. 6. Association rule {Internet Service = Fiber Optic, Tech Support = No, Online Security = No}→{Churn = Yes}.

4.2   Example - Independence

For this second example we investigate some variables that relate to *churn*. Other than the obvious variables like *tenure* and *contract* type, we also found that *online security* (Fig. 5a) and *tech support* (Fig. 5b) have a high impact on churn. In both cases not taking the service increases churn rate to around 42%. It is clear from the visualizations that these patterns have almost exactly the same distribution of customers. One would assume that these services are highly dependent, such that you can only enlist for security if you also take tech support and vice versa, as would be the case if they were part of a plan.

However by combining these variables in one visualization (Fig. 5c), we find that only about 72% ($\frac{0.36}{0.5}$) of the customers that don't have the security service also don't have tech support. Similarly the same relation holds for customers that don't have tech support. Again only about 72% of them didn't take the security service. In other words, the variables are less dependent than expected and hence the combination of the two also leads to a higher churn rate (50%) than either of them achieved independently (about 42% each).

Another unexpected variable that increases churn rate we found is *Fiber optic internet*. Using our visualization it is possible to play with these variables and create the desired association rules. For example an interesting task could be to find a subset of customers with a specific size, that has the highest chance of leaving. This would allow the telecom provider to invest its limited resources to counter churn in a more targeted strategy.

Figure 6 shows how this can be achieved with the help of our visualization. Items that increase churn can be added to the context, such that their interactions with other items can immediately be seen. At the same time the size of the target group and the churn rate remain visible. In this example we demonstrate a rule that selects 25% of the population with an elevated churn rate of 55%, which is quite remarkable considering that the most obvious and least influenceable attributes (*tenure* and *contract*) were not even included.

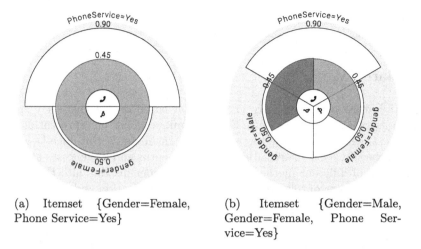

(a) Itemset {Gender=Female, Phone Service=Yes}

(b) Itemset {Gender=Male, Gender=Female, Phone Service=Yes}

Fig. 7. Two examples to illustrate the importance of taking context and meaning into account when evaluating patterns.

The telecom company can use this information to investigate why these unexpected variables have an impact on the churn rate. For example, offering free tech support might lead to less customers opting out and consequently to a potentially larger profit. This is however only speculation. An alternative explanation could be that customers who choose to pay for tech support tend to be prefer stability and are less prone to change between providers.

4.3   Example - Context

In this example, the importance of taking the context of a pattern into account is shown. Figure 7a displays what on first sight might appear to be a very interesting rule {*Gender = Female*} → {*Phone Service = Yes*} with a confidence of 90%. However with a closer inspection it becomes clear that 90% of the customers has *phone service*, independent of their gender, making this rule meaningless.

The second example (Fig. 7b) shows a pattern where an attribute is divided equally over two values. In a different context, this might be an interesting pattern. For example when comparing adolescents with the elderly, where we might *expect* a difference. In this case however, we know that phone service should not be biased towards a specific gender and we can deduce that it is not an interesting pattern.

Alternatively, one can sort or filter the association rules behind these patterns on how much their *lift* deviates from one. Since in this case the variables are independent, the association rules have a lift of one and would hence be filtered out or given a low rank. Under the expected independence assumption we would miss these patterns, where in fact they can be interesting when the domain implies an expected dependency. That is to say, both the context of a pattern and the larger context of the meaning of its attributes contribute to its *interestingness*.

5   Related Work

In the literature we can classify visualization based on whether they support itemsets, association rules or both. Some techniques focus on representing single itemsets [11] where others try to visualize the entire dataset at once [4,5,10,12, 13,16]. Naturally, each visualization has its strengths and weaknesses and hence repeated use of different methods can lead to deeper insights into the data [9]. We give an overview of related work in pattern visualization and remark on how these techniques relate to our visualization.

Single Itemset/Association Rule. Single Itemset/Association Rule Hofmann et al. discuss a double decker plot based on mosaic plots [11]. The idea is to visualize a single association rule with one item as consequent and provide metrics from which its interestingness can be assessed. In this visualization it is easy to verify that all items contribute to the rule, which likely indicates an interesting pattern. It is however rather limited in the amount of information that is visible or can be derived interactively. For example, because all possible subsets are rendered, the support also ends up scattered over the figure. The combination of a circular layout and our choice of subsets ensures that all segments are continuous in our visualization.

Circular. Two similar circular plots have been proposed earlier. However they differ from our visualization in that they both display the entire transaction database as a dissection in frequent itemsets. The first one by Dubois et al. called icVAT [4], has itemsets that radiate inward based on their support. Colors are used to show the cardinality of each itemset and the distance to the center (or radius) represents the support of each itemset.

In the second study by Keim et al. (FP-Viz [12]), items are layered to form itemsets and their support is indicated using a color scale and the width (or angle) of its segment. The main difference between these techniques is that icVAT has a fixed width and varying radius, where FP-Viz uses a fixed radius with varying width. Despite arguing that they make the link between items more clear, it remains difficult to see how different itemsets relate to each other since the same item can occur multiple times in different places, contrary to our technique.

Graph Based. Graph Based Ertek and Demiriz propose a straightforward graph based visualization [5]. Nodes represent items and special itemset nodes, with edges to the individual items, indicate itemsets. This distinction however makes it difficult to see interactions between itemsets and subsets, which is something our visualization excels in.

Leung et al. propose a different graph based method where nodes represent itemsets and they are organized according to their items and frequency [13]. This approach already depicts the interactions between itemsets and the differences in frequency more clearly. An issue with seeing the global picture remains however, since there is often no perceivable link between itemsets that share an item. In other words, it is difficult to trace itemsets to their sub- or supersets.

Bothorel et al. propose a graph based visualization with a circular layout to organize the nodes [2]. Itemsets are linked to their subsets and nodes that are linked also tend to be placed closer together. This visualization provides a good overview of what the data looks like and where the patterns may be, but patterns can no longer be discerned at a large scale.

Matrix/Table Based. Hahsler and Karpienko describe a grouped matrix representation in their study from 2017 [10]. This hierarchical approach enables the user to get an overview of the data at an abstract level and also to dive deeper to inspect more specific phenomena. Contrary to most techniques that try to visualize the entire dataset, this one is well equipped to handle the scaling problem. Its most prominent downside is that it projects antecedents and consequents on different axes, making it difficult to find interactions between them.

VisAR by Techapichetvanich and Datta [16] is a table based technique for visualizing association rules. It aims to provide a good overview that is efficient to query. Indeed, their technique lists association rules in a concise way with a clear and singular meaning. Since the visualization behaves like a table or list, it avoids screen clutter and occlusion. There is however no link between antecedent and consequent items in this visualization, which makes it hard to find interactions between rules. Furthermore long lists become impractical to use as well due to the increasing distance between items and between association rules.

6   Conclusions

A novel visualization technique for itemsets and association rules was introduced and analyzed from a visualization-theoretical perspective. Its functionalities and interactions were explained, making its application in data mining clear. The technique can be used as a concise representation for itemsets where the local and global context is immediately clear. In addition, the α-conditional view provides for an intuitive way to query interesting subsets of the data or to represent association rules. Furthermore the provided interactions allow an end user to actively query the visualization and extract valuable insights.

In the accompanying case study with a real-world dataset we demonstrated that a variety of patterns can be visualized and further understood with our technique. Users can combine their domain knowledge and expectations with the properties of itemsets and association rules to extract interesting information from the data. Furthermore, multiple instances of the visualization can be used together to describe complex relations.

Finally, we situated our technique in a broader context of itemset and association rule visualization techniques, remarking on differences and similarities. This thorough study of related work also supports our premise that the proposed visualization is effectively new in the field.

6.1 Future Work

We limited the features of our visualization to only the fundamental concepts of itemsets. No features for specific use cases were included, which of course leads to the benefit of making it usable in most contexts. However, as mentioned in Sect. 2.2, some degrees of freedom were left for annotation and integration. It would be interesting to see how the visualization can be expanded upon to provide functionality for specific use cases. For example in our case study with a clear target variable (*churn*), one could add a tooltip on hover that displays the fraction of items in the subset where *Churn = Yes*. If desired, this information could even be visualized by partially filling the segments with a fixed color.

Furthermore, to limit clutter we opted to keep the visualization sober and concise. More complex features could be integrated to wager this conciseness for more visible information. One idea is to also show relevant complementing itemsets, perhaps as an arc that radiates in from the outside. Similarly, additional arcs can be rendered to show the expected frequency of itemsets under certain independence assumptions, which would visualize *lift* more explicitly. Another potential addition would be to reserve space in the center of the circle for "common" items. These items could be selected interactively to limit the number of rendered itemsets to the ones containing these items, which would reduce occlusion. Finally, a user study could be performed to validate the effectiveness of our visualization.

Acknowledgements. This research received funding from the Flemish Government under the "Onderzoeksprogramma Artificiële Intelligentie (AI) Vlaanderen" programme.

References

1. Aggarwal, C.C., Bhuiyan, M.A., Hasan, M.A.: Frequent pattern mining algorithms: a survey. In: Aggarwal, C.C., Han, J. (eds.) Frequent Pattern Mining, pp. 19–64. Springer, Cham (2014). https://doi.org/10.1007/978-3-319-07821-2_2
2. Bothorel, G., Serrurier, M., Hurter, C.: Visualization of frequent itemsets with nested circular layout and bundling algorithm. In: Bebis, G., et al. (eds.) ISVC 2013. LNCS, vol. 8034, pp. 396–405. Springer, Heidelberg (2013). https://doi.org/10.1007/978-3-642-41939-3_38
3. Calders, T., Goethals, B.: Non-derivable itemset mining. Data Min. Knowl. Discov. **14**(1), 171–206 (2007)
4. Dubois, P.M., Han, Z., Jiang, F., Leung, C.K.: An interactive circular visual analytic tool for visualization of web data. In: 2016 IEEE/WIC/ACM International Conference on Web Intelligence (WI), pp. 709–712. IEEE (2016)

5. Ertek, G., Demiriz, A.: A framework for visualizing association mining results. In: Levi, A., Savaş, E., Yenigün, H., Balcısoy, S., Saygın, Y. (eds.) ISCIS 2006. LNCS, vol. 4263, pp. 593–602. Springer, Heidelberg (2006). https://doi.org/10.1007/11902140_63

6. Geng, L., Hamilton, H.J.: Interestingness measures for data mining: a survey. ACM Comput. Surv. (CSUR) **38**(3), 9 (2006)

7. Goethals, B.: Survey on frequent pattern mining. Univ. Helsinki **19**, 840–852 (2003)

8. Goethals, B., Moens, S., Vreeken, J.: MIME: a framework for interactive visual pattern mining. In: Proceedings of the 17th ACM SIGKDD International Conference on Knowledge Discovery and Data Mining, pp. 757–760. ACM (2011)

9. Hahsler, M.: arulesViz: interactive visualization of association rules with R. R J. **9**(2), 163–175 (2017). https://doi.org/10.32614/RJ-2017-047

10. Hahsler, M., Karpienko, R.: Visualizing association rules in hierarchical groups. J. Bus. Econ. **87**(3), 317–335 (2016). https://doi.org/10.1007/s11573-016-0822-8

11. Hofmann, H., Siebes, A.P., Wilhelm, A.F.: Visualizing association rules with interactive mosaic plots. In: Proceedings of the sixth ACM SIGKDD International Conference on Knowledge Discovery and Data Mining, pp. 227–235. ACM (2000)

12. Keim, D.A., Schneidewind, J., Sips, M.: FP-Viz: visual frequent pattern mining. In: InfoVis (2005)

13. Leung, C.K.-S., Irani, P.P., Carmichael, C.L.: FIsViz: a frequent itemset visualizer. In: Washio, T., Suzuki, E., Ting, K.M., Inokuchi, A. (eds.) PAKDD 2008. LNCS (LNAI), vol. 5012, pp. 644–652. Springer, Heidelberg (2008). https://doi.org/10.1007/978-3-540-68125-0_60

14. Moody, D.: The "physics" of notations: toward a scientific basis for constructing visual notations in software engineering. IEEE Trans. Softw. Eng. **35**(6), 756–779 (2009)

15. Tan, P.N., Kumar, V., Srivastava, J.: Selecting the right interestingness measure for association patterns. In: Proceedings of the Eighth ACM SIGKDD International Conference on Knowledge Discovery and Data Mining, pp. 32–41. ACM (2002)

16. Techapichetvanich, K., Datta, A.: VisAR : a new technique for visualizing mined association rules. In: Li, X., Wang, S., Dong, Z.Y. (eds.) ADMA 2005. LNCS (LNAI), vol. 3584, pp. 88–95. Springer, Heidelberg (2005). https://doi.org/10.1007/11527503_12

17. Vreeken, J., Van Leeuwen, M., Siebes, A.: KRIMP: mining itemsets that compress. Data Min. Knowl. Disc. **23**(1), 169–214 (2011)

18. Yamamoto, C.H., Oliveira, M.C.F., Rezende, S.O.: Including the user in the knowledge discovery loop: interactive itemset-driven rule extraction. In: Proceedings of the 2008 ACM Symposium on Applied Computing, pp. 1212–1217 (2008)

Understanding Telecom Customer Churn with Machine Learning: From Prediction to Causal Inference

Théo Verhelst[1]([✉]), Olivier Caelen[2], Jean-Christophe Dewitte[2], Bertrand Lebichot[1], and Gianluca Bontempi[1]

[1] Computer Science Department, Machine Learning Group, Université Libre de Bruxelles, Brussels, Belgium
{tverhels,blebicho,gbonte}@ulb.ac.be
[2] Data Science Team, Orange Belgium, Brussels, Belgium
{olivier.caelen,jean-christophe.dewitte}@orange.be

Abstract. Telecommunication companies are evolving in a highly competitive market where attracting new customers is much more expensive than retaining existing ones. Though retention campaigns may be used to prevent customer churn, their success depends on the availability of accurate prediction models. Churn prediction is notoriously a difficult problem because of the large amount of data, non-linearity, imbalance and low separability between the classes of churners and non-churners. In this paper, we discuss a real case of churn prediction based on Orange Belgium customer data. In the first part of the paper we focus on the design of an accurate prediction model. The large class imbalance between the two classes is handled with the EasyEnsemble algorithm using a random forest classifier. We assess also the impact of different data preprocessing techniques including feature selection and engineering. Results show that feature selection can be used to reduce computation time and memory requirements, though engineering variables does not necessarily improve performance. In the second part of the paper we explore the application of data-driven causal inference, which aims to infer causal relationships between variables from observational data. We conclude that the bill shock and the wrong tariff plan positioning are putative causes of churn. This is supported by the prior knowledge of experts at Orange Belgium. Finally, we present a novel method to evaluate, in terms of the direction and magnitude, the impact of causally relevant variables on churn, making the assumption of no confounding factors.

Keywords: Churn prediction · Machine learning · Big data · Causal inference

1 Introduction

In recent years, the number of mobile phone users had a massive increase, reaching more than 3 billion users worldwide. The number of mobile phone service subscriptions is greater than the number of residents in several countries, including

© Springer Nature Switzerland AG 2020
B. Bogaerts et al. (Eds.): BNAIC 2019/BENELEARN 2019, CCIS 1196, pp. 182–200, 2020.
https://doi.org/10.1007/978-3-030-65154-1_11

Belgium [13]. Telecommunication companies are evolving in a saturated market, where customers are exposed to competitive offers from many other companies. Hadden et al. [11] showed that attracting new customers can be up to six times more expensive than retaining existing ones. This led companies to switch from a sale-oriented to a customer-oriented marketing approach. By building customer relationships based on trustworthiness and commitment, a telecommunication company can reduce churn, therefore increasing benefits through the subsequent customer lifetime value. A typical marketing strategy to improve customer relationship is to conduct retention campaigns whose effectiveness depends on accurate profiles of customers (e.g. in terms of attrition risk).

Churn detection is nowadays performed by most major telecommunication companies using machine learning and data mining [5,12,18,28–31]. Churn prediction is a notoriously difficult learning task because of the large quantity of data, non-linearity, imbalance and low separability between the classes of churners and non-churners. The first part of the paper assesses several machine learning methods and strategies by using a large dataset measuring the churn behavior of Orange Belgium telecom clients. Estimating the probability of churn of a customer is however not sufficient if we wish to design an effective retention campaign (e.g. based on incentives). For this reason, the second part of the paper explores the adoption of causal techniques to infer from observational data the most probable causes of a churn behavior. Causal analysis is usually conducted through *controlled randomized experiments* [7], by evaluating the impact of a potentially causal variable on the target variable. In the context of customer relationship management, controlled experiments are possible through the retention campaigns, where the offers made to the customers act as variable manipulations. Though this reduces the risk of confounding factors, access to such data is typically difficult and expensive. For this reason, we have recourse to data-driven inference approaches, which aim to reconstruct causal dependencies based on the statistical distribution of the considered variables. Most existing approaches however make different assumptions about the data distributions which are difficult to assess in practice. For this reason, we adopt a "wisdom of the crowd" approach by running in parallel several state-of-the-art approaches and combining their results for final considerations. Also to assess the quality of the obtained putative causes we estimate from data the causal impact of every single cause on churn probability.

We may summarize the main contributions of the article as follows.

– Assessment of a state-of-the-art churn prediction pipeline and study of the impact of several model variants (e.g different feature sets and different subscription contracts) (Sect. 2).
– Application of causal strategies to infer putative causes of churn from observational data (Sect. 3).
– Assessment of the impact of putative causal variables on churn (Sect. 3).

The rest of this paper is structured as follows. In Sect. 2 we describe the dataset, the machine learning pipeline and the results of churn prediction. In Sect. 3 we provide a causal analysis of churn. Conclusion and future work perspectives are discussed in Sect. 4.

2   Churn Prediction

This section describes the Orange dataset and the machine learning pipeline designed to assess a number of strategies and models for predicting the probability of customer churn.

2.1   Data

The dataset is a monthly report of Orange Belgium customers' activity covering a 5 months time window in 2018. For confidentiality reasons, we will disclose here only some high-level details about the dataset. The dataset contains 73 features about customer activity including subscription type, used hardware, mobile data usage (in MB), number of calls/messages and some socio-demographic information. The dataset has 5.3 million entries (about 1 million entries per month). The target variable, denoting churn, is binary and takes the true value if the client is known to have churned in the two months following the input timestamp. The churn prediction problem is highly unbalanced, since there are far more non-churners than churners.

Two kinds of subscriptions are present in this dataset: SIM-only[1] and loyalty. The first refers to a subscription where the customer can quit at any time with no cost. In the loyalty contract the customer is rewarded (e.g. discount on the purchase of a mobile phone) in exchange of a fixed contract duration (e.g. 24 months). If the customer decides nonetheless to stop his subscription before the term of the contract, he has to pay back the reward. There are about 5 million entries in the SIM-only dataset, and about 250,000 entries in the loyalty dataset. In this paper, we will mainly focus on SIM-only contracts, given its broader impact on the Orange customer base and the larger statistical power due to the availability of more samples. Some experiments have been conducted anyway on both contract types, to understand the differences in terms of churn behavior.

In order to provide a visual description of the informative content of the dataset, let us consider in Fig. 1 two variables having a clear relation with churn. The horizontal axis indicates whether a customer has a cable connection while the vertical axis denotes the payment responsible (taking a "No" value when someone else than the customer, e.g. a parent, pays the bill). It appears that most Orange Belgium customers do not have a cable connection and are responsible for the payment. The color of the spots indicates the churn rate, with a lighter color denoting a higher probability of churn[2]. The impact of both binary

---

[1] *SIM-only* indicates that the customer bought no other product than the SIM card.

[2] For confidentiality reasons, the precise value of the churn rate cannot be disclosed.

variables appears clearly, with a significant difference in churn rate between the two extremes. The univariate impact of each variable on churn can be quantified in terms of odds ratio, measuring the increase of the odds of churn once exposed (i.e. when the customer is responsible for the payment, or when there is a cable connection). The odd ratios for the payment responsible and the cable connection are 0.917 and 0.839, respectively. This indicates that a "Yes" value for both variables is associated to a reduced risk of churn.

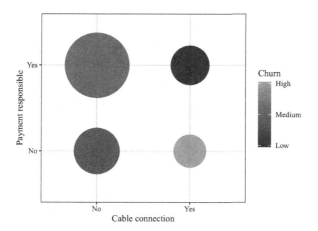

Fig. 1. Interaction between cable connection and payment responsible. A customer is not responsible for payment if someone else (e.g. a parent) pays the invoice in her stead. The color of the spots denotes the churn rate, whereas its area denotes the number of customers.

Another interesting visualization concerns the relation between tenure (i.e. the duration of the current subscription) and churn rate (Fig. 2)[3]. The curve shows a negative correlation between the churn rate and tenure. Note that the surge in churn rate corresponds to the term of the contract for loyalty customers.

2.2   Machine Learning Pipeline

Three different learning tasks are created by stratifying the dataset: one containing the loyalty contracts, one containing the SIM-only contracts, and one containing the SIM-only contracts with additional variables (denoted SIM-only Δ). The large unbalancedness of the dataset has been addressed by adopting the EasyEnsemble strategy [16] which consists in training a number (in our case 10) of learners on the whole set of positive instances (churners) and an equally sized random set of negative instances. Based on our previous experience on related

---

[3] For confidentiality reasons, the axes scales are concealed.

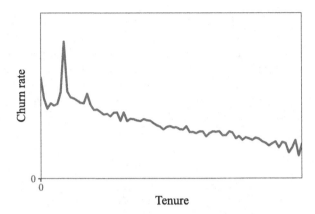

Fig. 2. Churn rate as a function of the tenure (the duration of the current subscription). The spike on the left of the plot corresponds to the end of the loyalty period for loyalty customers.

largely unbalanced tasks (notably fraud detection [3,4]) we considered as learner only Random Forests.

In what follows we report the results of a number of assessments evaluating the impact of

1. variable selection, based on the feature importance returned by Random Forest;
2. the addition of engineered features: for each time-dependent quantity (e.g. total duration of calls, or mobile data usage) we created 2 additional features measuring the difference and the ratio between two consequent monthly values, respectively;
3. the type of contract (SIM-only vs. loyalty).

The high computational cost of training on such a large dataset restricts the number of configurations we can assess. We limit the number of selected variables to 20, 30 or all variables. Also, we do not explore the difference variables for loyalty contracts. Overall we consider 9 different experiment configurations.

Three-fold cross-validation is used to assess the accuracy on the training set (first 4 months). The last month of data is used as a test set for each of the three datasets, in order to check the robustness of the prediction model (e.g. with respect to potential drifts or non-stationarity).

The performance of the different models is evaluated using three different measures: the receiver operating characteristic (ROC) curve, the precision-recall (PR) curve, and the lift curve [27]. While the ROC curve and the PR curve are widely used in conventional classification, the lift curve is of more practical interest in evaluating churn prediction. Since a customer churn retention campaign focuses on a limited amount of customers, the lift curve reports the expected performance of the model as the number of customers included in the campaign varies. From these curves, we derive the area under the ROC curve

(AUROC), the area under the PR curve (AUPRC) and the lift at different thresholds (1%, 5%, and 10% of customers included). There is some evidence in the literature [6,9,22] that the ROC curve is not a reliable metric on unbalanced data. Moreover, since the area under the ROC curve depends on all possible decision thresholds, it does not correspond with the objective of the churn campaign: finding a small group of customers with high churn probability (low false-positive rate). We report the AUROC to be consistent with the churn prediction literature, but our conclusions are mainly based on the other performance metrics.

2.3   Results and Discussion

Table 1 and 2 report the cross-validation and the test accuracy, respectively. Based on those results, a number of considerations can be made

- by reducing the number of features to 30, the accuracy does not deteriorate significantly. This is good news for our industrial partner since a compact churn model is more suitable for production.
- though adding engineered features may be beneficial, this occurs only if a feature selection is conducted beforehand.
- surprisingly, the accuracy is higher for the test set (Table 2) than in cross-validation (Table 1). Our interpretation, confirmed by visualization in the space of the two first principal components, is that the drift of the data makes the classification easier.
- regarding the type of contracts, churn is slightly easier to predict in the loyalty dataset than SIM-only, due to the greater importance of time-related variables. Indeed, the churn is significantly higher at the end of the mandatory period of a loyalty contract, facilitating the prediction process.

We compared our results on the SIM-only dataset with other published studies on churn prediction [5,12,18,28–31]. We achieve similar results in terms of area under the ROC curve and lift.

Table 1. Summary of the cross-validation results. Highest values for each type of contract and for each evaluation measure are underlined.

	SIM-only			SIM-onlyΔ			Loyalty		
	20	30	All	20	30	All	20	30	All
AUROC	0.64	0.73	0.74	0.74	0.74	0.70	0.76	0.78	0.77
AUPRC	0.04	0.08	0.08	0.09	0.09	0.07	0.13	0.16	0.15
Lift at 10%	2.10	3.16	3.39	3.39	3.44	3.01	3.22	3.57	3.50
Lift at 5%	2.41	4.11	4.52	4.49	4.57	3.90	3.71	4.30	4.18
Lift at 1%	3.24	7.58	8.36	8.80	8.67	6.79	5.00	6.37	6.11

Table 2. Summary of the results of prediction experiments on the test set. Highest values for each type of contract and for each evaluation measure are underlined. Using only 20 variables decreases the performances most often.

	SIM-only			SIM-only Δ			Loyalty		
	20	30	All	20	30	All	20	30	All
AUROC	0.66	<u>0.73</u>	<u>0.73</u>	0.72	<u>0.73</u>	0.69	0.74	<u>0.76</u>	<u>0.76</u>
AUPRC	0.05	<u>0.10</u>	<u>0.10</u>	<u>0.10</u>	<u>0.10</u>	0.08	0.15	<u>0.19</u>	0.18
Lift at 10%	2.25	3.34	3.41	3.27	<u>3.42</u>	3.03	2.96	<u>3.40</u>	3.30
Lift at 5%	2.64	4.49	<u>4.68</u>	4.48	4.67	4.09	3.51	<u>4.22</u>	4.02
Lift at 1%	4.29	9.20	9.53	<u>10.09</u>	9.95	7.67	4.66	<u>6.65</u>	6.16

3   Causal Analysis

The selection procedure discussed in the previous section returns the most relevant variables for predicting the potential churners. Though such variables provide useful information for designing a predictor, they are not necessarily the ones to manipulate (e.g. by giving incentive) if we wish to reduce the churn risk. For example, an increase in the number of contracts registered by a customer may be strongly associated with a decrease of churn. However, a hypothetical churn retention action that would sell additional contracts might fail, if customer satisfaction has a causal effect both on the number of purchased contracts and the propensity to churn. In this case, the predictive variable (number of contracts) and the churn have a common latent cause (customer satisfaction). Manipulating the number of contracts will therefore not affect churn. Different tools are needed to discover true causal relationships between variables and will be discussed in what follows.

3.1   Causal Inference Strategy

We use the dataset of Sect. 2 and perform the causal inference separately on the SIM-only and loyalty customers since it is supposed that the causes of churn are at least partially different between loyalty and SIM-only contracts. All 5 months of data are used. To decrease computation time, only the first 30 variables in the ranking of the random forest trained in Sect. 2 are used. A random undersampling has been applied to reach decent computation times and to perform class balancing. The positive class (churners) is kept fixed and a random subset of the negative class is randomly selected, such that the resulting dataset contains the same number of positive and negative observations. Further random undersampling is then performed, with a sample size depending on the inference algorithm.

The rationale of this experiment consists in applying several causal inference techniques, which give different types of results in various forms, and extract a consensus, if any, in the light of the different assumptions each model puts on the data. This strategy is called triangulating [14] and takes advantage of the fact that the different causal inference methods rely on different assumptions, thus increasing the validity of our results.

The considered causal inference algorithms are: PC [24], Grow-shrink (GS) [17], Incremental Association Markov Blanket (IAMB) [26], minimum Interaction Maximum Relevance (mIMR) [2] and D2C [1]. PC infers the equivalence class of causal graphs faithful to the dataset, GS and IAMB infer the Markov blanket of the churn variable, and mIMR and D2C return the direct causes of churn.

The PC [24] algorithm is slow when the number of samples is large since the whole causal network is inferred. Therefore, we restrict the dataset to 10,000 samples for this algorithm. The result is given under the form of a completed partial DAG (CPDAG) representing an equivalence class of directed acyclic graphs (DAG) [25].

The GS and IAMB algorithms [17,26] both infer the Markov blanket of a target variable, the churn in our case. These algorithms therefore return the direct causes, the direct effects and the direct causes of the direct effects (also called *spouses*). For these two algorithms, the entire set of positive samples is used, along with a subset of the same size of negative samples. IAMB differs from GS in that it is more sample-efficient.

Two implementations of the mIMR algorithm [2] are used: one based on histograms to estimate mutual information, and another assuming Gaussian distributions, thus allowing a closed-form formula for the computation of the mutual information [19]. For the first implementation, the dataset is restricted to 10,000 samples, due to the computational cost of the histogram-based estimator. In the second implementation, 100,000 samples from SIM-only are used, and all samples from loyalty are used. The results are provided as a list of the first 15 selected variables, accompanied by the gain provided by each variable at each iteration of the algorithm.

The D2C learning algorithm is trained using randomly generated DAGs, as described in [1]. We assume a Markov blanket of 4 variables when constructing the asymmetrical features. Given the high computational cost of feature extraction, 2,000 samples are used from the customer dataset. The results are provided as the predicted probability for each variable to be a cause of churn.

For the first three methods (PC, IAMB, and GS), we use the R package *bnlearn* [23] for independence tests using mutual information and asymptotic χ^2 test [8]. For mIMR and D2C, we use the R package *D2C* [1]. In all cases, a false-positive rate of 0.05 is chosen for statistical tests of independence.

Before proceeding with the results, it is worth reminding that all the 5 causal inference algorithms rely on specific assumptions. While PC, GS, and IAMB assume causal sufficiency and faithfulness, mIMR and D2C rely on more specific conditions.

Causal sufficiency denotes the absence of unmeasured confounder, and is likely to hold given the large number of variables (73) and the variety of information they provide (service usage, socio-demographic, type of subscription, etc). Confounding could be further reduced by including an indicator of service quality, which is absent from our dataset. See Sect. 3.3 for a more detailed review of our prior knowledge on the causes of churn.

The assumption of faithfulness states that any (conditional) independence found in the probability distribution is reflected by the d-separation of the relevant variables in the corresponding causal graph. Faithfulness in the case of the PC algorithm is discussed in [15].

mIMR is based on the assumption that direct causes form "unshielded collider" configurations together with the target. Since in such configurations direct causes are marginally independent and conditionally dependent, mIMR may exploit this pattern to prioritize direct classes in the ranking. Though such an assumption is hardly satisfied in real settings, the approach allows to introduce a causal criterion in a feature selection algorithm for large dimensional settings. The adoption of mIMR requires as well the choice of a mutual information estimator (typically Gaussian for its low-variance and robustness in non-normal configurations [19]).

D2C relies on the existence of asymmetric descriptors of the statistic dependency between a cause/effect pair. This is possible under some specific conditions, like the existence of a single edge connecting the Markov blanket of the cause and the effect and the existence of an unshielded collider between the cause (effect) and the related spouse. While the first assumption is probably not true in our setting, the second is satisfied by the fact that no descendant of the target variable is included in the dataset.

3.2   Sensitivity Analysis

Once causally relevant variables are inferred, it is worth evaluating the sensitivity of the target to their manipulation. In Sect. 2 we learned a predictive algorithm (random forest) to estimate $P(Y \mid X)$, i.e. the conditional probability distribution of the churn variable Y given the set of customer variables $\{X_1, \ldots, X_n\} = X$. Let us now focus on a putative cause $X_i \in X$ and assume causal sufficiency, i.e. all possible confounders are part of the set of measured variables X. In order to assess the sensitivity of Y to X_i, we measure the change in $P(Y \mid X)$ once the distribution of X_i is modified. This boils down to estimate the *natural direct effect* [21], which quantifies the causal effect of X_i on Y not mediated by any other variable, while the other variables are still distributed according to their natural distribution. This corresponds to answering the causal question "*What happens if only X_i changes?*".

Since we are interested in the effect of a shift in the distribution of X_i, we add $\alpha \sigma_i$ to the value of X_i, where σ_i is the standard deviation of X_i and α is a parameter of the intervention. The natural direct effect is

$$\mathrm{NDE}_{x_i} = \mathrm{NDE}_{x_i}(Y) - E[Y]$$

where $\mathrm{NDE}_{x_i}(Y)$ is defined as the expected value of Y under "natural" intervention on X_i, i.e. by letting other variables be distributed according to their original distribution:

$$\mathrm{NDE}_{x_i}(Y) = \int P(x)P(Y = 1 \mid x_1, \ldots, \mathrm{do}(x_i + \alpha\sigma_i), \ldots, x_n) \, dx.$$

We know that all possible back-door paths between X_i and Y are blocked since, by causal sufficiency, $X = \{X_1, \ldots, X_n\}$ includes all possible confounders. Therefore, using rule 2 of do-calculus [20] we can estimate $\mathrm{NDE}_{x_i}(Y)$ from observational data alone.

Given a dataset of n variables and N examples $\{(x_1^{(j)}, \ldots, x_n^{(j)}; y^{(j)})\}_{1 \leq j \leq N}$, the average prediction of a model f on this dataset is an estimator of the expected value of Y:

$$E[Y] \approx \frac{1}{N} \sum_{j=1}^{N} f(x_1^{(j)}, \ldots, x_n^{(j)})$$

For each variable X_i the expected value of Y under intervention can be approximated as

$$\mathrm{NDE}_{x_i}(Y) \approx \frac{1}{N} \sum_{j=1}^{N} f(x_1^{(j)}, \ldots, x_i^{(j)} + \alpha\sigma_i, \ldots, x_n^{(j)})$$

We applied this method on the SIM-only dataset, on the 30 variables having the largest importance according to the random forest models. The causal effect is computed for $\alpha = 1$ and $\alpha = -1$. The assumption of causal sufficiency (Sect. 3.1) is a necessary condition for the validity of this method. Note that the dataset we use in practice also contains discrete variables. These variables are left out of this analysis since the method is suited only to continuous variables.

3.3   Prior Knowledge

Before presenting the results of causal inference, it is interesting to summarize the knowledge of the Orange experts on the possible reasons for customer churn, elicited by means of several discussions and interviews. Those experts report four main causes of churn:

Bill shock: this occurs when a customer has an unusually large service usage, which results in an important "out of bundle" amount (i.e. the client is charged much more than usual). This triggers a reaction from the customer inducing an increased risk of churn. This scenario is well understood and verified in practice. It is believed to be the most important cause of churn.

Customer dissatisfaction: multiple factors influence customer satisfaction, including quality of service and network quality. A customer having numerous cuts of network connection during phone calls, or unable to use properly Orange online services, will be more likely to seek better alternatives elsewhere.

Wrong positioning: choosing the right tariff plan suited to one's service usage habits is sometimes difficult. On the one hand, if not enough call time is provisioned, an "out of bundle" amount is likely to be charged at the end of the month. On the other hand, an expensive tariff plan results in a high fixed cost for the customer. When the needs of a customer do not correspond to the chosen tariff plan, we say that the customer is wrongly positioned. A wrong positioning results in most cases to a higher bill than expected, and is a significant cause of churn.

Churn due to a move: it is common to choose a product bundle from a telecommunication company comprising a subscription for mobile phone, landline phone, television, and internet connection. In this case, the subscription is tied to the particular place of domicile of the customer. When the client moves to another place, it is quite common to also change for another telecommunication service provider. Therefore, this is a significant cause of churn, albeit of a different nature from the other settings exposed above.

While these potential causes of churn pertain to the whole customer base, the loyalty customers typically have a much higher churn rate at the end of the mandatory period of their contract, thus the tenure (the time since when a customer uses Orange' services) is an important cause of churn for them. On SIM-only customers, expert knowledge also indicates that the tenure influences churn: a new customer is more likely to churn than a long-time customer since it is less committed to the company.

These different settings are described informally, and their translation to the formal definitions of causality is not straightforward. We wish to find a mapping between the events believed to be causes of churn and specific instantiations of measurable random variables. In the case of the first setting (bill shock), we can reasonably assume that variables measuring the "out of bundle" amount of the customer is a faithful proxy for bill shock. Similarly, customer satisfaction can be estimated using, for example, the number of network cuts during phone calls, or the number of calls to the customer service. The wrong positioning can also be numerically estimated, given the tariff plan of the client and its average service usage. The last setting (churn due to a move) is much more difficult to account for, as it is not directly related to the interaction between the client and the telecommunication services.

In the dataset available for this study, the only measured variables that translate to potential causes of churn are the "out of bundle", the tariff plan and service usage (phone calls, messages, mobile data). We have no measure for network quality, customer satisfaction, or propensity to move soon. Also, the wrong positioning is not explicitly encoded and has to be inferred by the causal inference model from the average service usage and the current tariff plan.

3.4   Results of Causal Inference

The outcome of the inference algorithms is summarized in Figs. 3 and 4. Each of the possible causes of churn is represented by an ellipse, annotated with the algorithms that output this variable. For both SIM-only and loyalty, the PC algorithm infers an intricate causal graph but where the churn variable is disconnected from all others. Note that GS and IAMB output the Markov blanket, and not only direct causes. Since the output of mIMR is a ranking, we use background knowledge to determine how many of the top-ranked variables should be considered as inferred causes, based on their redundancy. In the case of the histogram-based mIMR, the first variables in the ranking are complementary, but the 10th variable (for SIM-only) and the 12th variable (for loyalty) are mostly redundant with the other variables higher in the ranking. This indicates that the variable interaction is low and the remaining variables lower in the ranking should not be considered as causes. For the mIMR with Gaussian assumption, there is a significant drop in the gain between the 7th and the 8th ranked variables in the SIM-only dataset, and between the 8th and 9th ranked variables in the loyalty dataset. We consider that as a criterion for considering only the 7 (SIM-only) and 8 (loyalty) first ranked variables as inferred causes. D2C outputs a probability of being a cause of churn, for each variable. For the SIM-only dataset, we selected the tariff plan, the province of residence and the data usage as causes inferred by D2C, and for the loyalty dataset, we selected the number of contracts, the province, and the tenure. These variables display a significantly higher predicted score than the other variables.

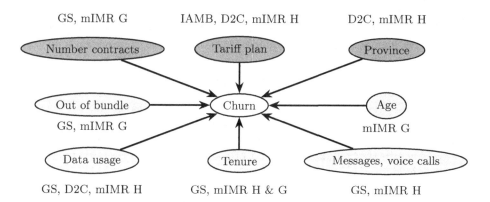

Fig. 3. Summary of results of causal inference on SIM-only dataset. Each variable is annotated with the algorithms predicting it to be a cause of churn. Light gray ellipses represent continuous variables, and dark gray ellipses represent discrete variables. mIMR H stands for the histogram-based estimator, and mIMR G for the estimator with Gaussian assumption.

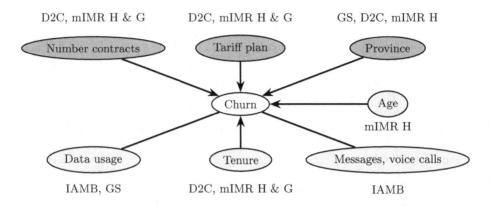

Fig. 4. Summary of results of causal inference on loyalty dataset. Each variable is annotated with the algorithms predicting it to be a cause of churn. Light gray ellipses represent continuous variables, and dark gray ellipses represent discrete variables. mIMR H stands for the histogram-based estimator, and mIMR G for the estimator with Gaussian assumption.

For the SIM-only dataset (Fig. 3), the "out of bundle" and data usage variables are reported as causes by mIMR and D2C, and as members of the Markov blanket by GS. This is in line with our prior belief that bill shock is a major cause of churn. We could expect the "out of bundle" variable to stand out more explicitly, but it is only given by mIMR with Gaussian assumption. However, the distribution of the "out of bundle" can roughly be modeled as the exponential of a Gaussian. It is thus easy to understand why the other inference methods, that make different statistical assumptions, fail to report the causal link to churn.

The tariff plan and the "out-of-bundle" variables together provide a representation of the tariff plan positioning of the customer. For the SIM-only dataset, these two variables are reported as causes of churn by mIMR and D2C and are also members of the Markov blanket according to GS and IAMB. This confirms our hypothesis that wrong positioning is an important cause of churn.

Note that the "out of bundle" is not reported by any algorithm for the loyalty dataset (Fig. 4). This is consistent with the fact that loyalty customers are not able to churn in the mandatory period of their contract, thus churn related to bill shock is less represented in this dataset.

The two last causes of churn according to Sect. 3.3 are customer satisfaction and churn due to a move. None of the measured variables are direct proxies for these two putative explanations of churn. Better results could be obtained by using relevant variables such as, for example, the number of calls to the customer service, a measure of the network quality, the number of network cuts during a call, and so on. Adding these variables would reduce latent confounding if the underlying causal hypotheses are true. We suspect that the importance of the province in Figs. 3 and 4 is an indication that network quality is an important cause of churn (the network quality is known to vary between different regions of Belgium). However, the scope of this study limited us to the set of variables presented in Sect. 2.1.

If we use the expert knowledge to assess the accuracy of the causal inference algorithms, mIMR H and D2C algorithms seem to better infer relevant variables as direct causes. Indeed, the bill shock and the wrong positioning imply that the "out of bundle", the tariff plan and the data usage are likely causes of churn. The two latter are output by mIMR H and D2C in the SIM-only dataset, whereas mIMR G outputs the "out of bundle". In the loyalty dataset, D2C and mIMR correctly avoid reporting the "out of bundle" or the data usage as causes of churn, but correctly report the importance of the tenure. A model similar to mIMR H or D2C, but able to correctly handle variables with more difficult distributions such as the "out of bundle" variable, would be ideal.

Finally, it is important to consider that these results may suffer from sampling bias. Given that we use a crude random undersampling technique, some causal patterns in the discarded positive samples may be under-represented in the resulting training set. This is especially the case for the PC algorithm (using 10,000 samples), the first implementation of mIMR (10,000 samples), and D2C (2,000 samples). And even though the remaining algorithms use far more samples, none of them can take into account the entire set of non-churners. Furthermore, we have no theoretical guarantee that an even class ratio is best for inferring causal patterns. Reducing sampling bias in causal analysis requires the conception of new techniques that are outside the scope of this article.

3.5 Results of Sensitivity Analysis

The results of the variable sensitivity analysis are shown in Figs. 5 and 6. Each variable is represented as a bar whose color depends on the category of variable: subscription, calls and messages, mobile data usage, revenue, customer hardware, and socio-demographic. Some variable names have been anonymized for confidentiality reasons. Also, only variables inducing the most significant change in the distribution are shown ($p < 10^{-10}$ with a two-sided t-test).

All the numerical variables inferred as possible causes of churn appear to influence the predictions of the model, albeit in a non-linear manner as indicated by the lack of symmetry between Figs. 5 and 6. On the one hand, the tenure and the number of contracts are observed to be monotonically associated with the churn probability, since they appear in both figures in opposite directions. On the other hand, variables related to the amount paid by the customer and the data usage cause more churn when they are increased, but the opposite is not true. Note that the tariff plan and the province, although reported as possible causes in Fig. 3, are not present in Figs. 5 and 6 since they are categorical, thus unsuitable for this analysis.

In Appendix A (Figs. 7 and 8), we report the entire distribution of predicted probability of churn for each shifted variable, instead of reporting only the difference between the means as in Fig. 6 and 5. This shows other characteristics of the causal impact of these variables on churn, such as the change in the spread of the probability distribution. Note that, while the churn is highly unbalanced in the original dataset, the predicted probability of churn is balanced. This is due to the EasyEnsemble methodology, which generates balanced subsets of the original training set.

The causal impact of a smaller intervention is reported in Appendix A, Figs. 9 and 10. The intervention consists in adding or subtracting $0.5\,\sigma_i$ from each variable separately, instead of σ_i as in Figs. 5, 6, 7, and 8. We observe that some variables have almost the same impact as with a shift of 1 sigma (e.g. the out of bundle variables), while others have significantly less impact, such as the number of contracts. In the latter case, this is due to the discrete distribution of the number of contracts. Other variables, such as D13, have a proportionally reduced impact on the predicted probability of churn.

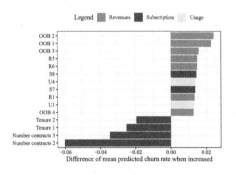

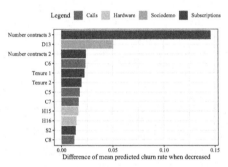

Fig. 5. Difference of mean predicted probability of churn when a standard deviation is added separately to each variable. Run on the SIM-only dataset. Only variables inducing the most significant change in the distribution are shown ($p < 10^{-10}$ with a two-sided t-test).

Fig. 6. Difference of mean predicted probability of churn when a standard deviation is subtracted separately from each variable. Run on the SIM-only dataset. Only variables inducing the most significant change in the distribution are shown ($p < 10^{-10}$ with a two-sided t-test).

4 Conclusion

Churn prediction in the telecommunication industry is notoriously a hard task characterized by the non-linearity of variables, large overlap between churners and non-churners, and class imbalance. Predictive modeling of churn was achieved with a random forest classifier and the Easy Ensemble algorithm. In a series of experiments on churn prediction, we assessed the impact of variable selection, type of contract and use of engineered features. The results show that variable selection helps reducing computation time if at least 30 features are selected. Also, the engineering of new features may be beneficial if variable selection is applied. We explored the application of causal inference from observational data. More specifically, we applied 5 different causal inference methods, namely PC, Grow-Shrink (GS), Incremental Association Markov Blanket (IAMB), minimum Interaction Maximum Relevance (mRMR), and D2C. The results of these algorithms are heterogeneous yet consistent with prior knowledge of the causes of churn. The direction of the causal influence of variables on churn was estimated through a novel method of sensitivity analysis. This method is based on the assumption that no latent variables are confounding factors of churn and the variable under inspection. This method showed that some variables have a non-monotonic causal influence on churn, which is consistent with expert knowledge.

5   Future Work

Results of causal analyses are difficult to validate without the ability to perform experiments. In this study, we are limited to compare our findings with prior knowledge of experts. Retention campaigns provide a promising opportunity to validate causal hypotheses. They can emulate a variable manipulation by offering risky customers targeted promotions. We plan to conduct such experiments in the future through collaboration with Orange Belgium.

Uplift modeling is an interesting approach to incorporate causal consideration in churn prediction [10]. In uplift modeling, the customers are ranked according to the diminution of their probability of churn when subject to the campaign, as opposed to usual churn modeling that ranks customers according to their probability of churn. Retention campaigns will allow assessing the effectiveness of this approach.

Another limitation of our approach is the arbitrary decision threshold we fixed between inferred causes and non-causes for the mIMR and D2C algorithms. Since these two methods output a score for each variable, we can instead compute a performance curve (*e.g.* ROC, precision-recall) from the predicted scores and the ground truth provided by experts. Although this is not suitable for performing causal discovery *per se*, this allows to quantitatively evaluate causal inference algorithms.

Undersampling and class balancing are used to ensure the computational tractability of causal inference. However, this may result in sampling bias, and its effect on the results has not been formally assessed. We can obtain more robust and stable results by performing undersampling and the causal inference experiments multiple times, as in the EasyEnsemble methodology.

A   Additional Figures on Sensitivity Analysis

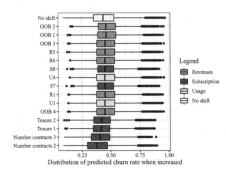

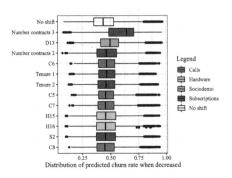

Fig. 7. Distribution of the predicted probability of churn when a standard deviation is added separately to each variable. Run on the SIM-only dataset. Only variables inducing the most significant change in the distribution are shown ($p < 10^{-10}$ with a two-sided t-test).

Fig. 8. Distribution of the predicted probability of churn when a standard deviation is subtracted separately from each variable. Run on the SIM-only dataset. Only variables inducing the most significant change in the distribution are shown ($p < 10^{-10}$ with a two-sided t-test).

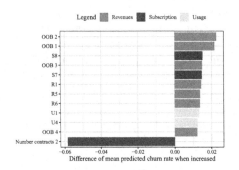

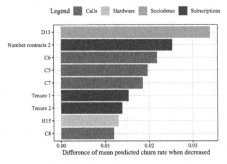

Fig. 9. Difference of mean predicted probability of churn when half a standard deviation is added separately to each variable. Run on the SIM-only dataset. Only variables inducing the most significant change in the distribution are shown ($p < 10^{-10}$ with a two-sided t-test).

Fig. 10. Difference of mean predicted probability of churn when half a standard deviation is subtracted separately from each variable. Run on the SIM-only dataset. Only variables inducing the most significant change in the distribution are shown ($p < 10^{-10}$ with a two-sided t-test).

References

1. Bontempi, G., Flauder, M.: From dependency to causality: a machine learning approach. J. Mach. Learn. Res. **16**(1), 2437–2457 (2015)
2. Bontempi, G., Meyer, P.E.: Causal filter selection in microarray data. In: Proceedings of the 27th International Conference on Machine Learning (icml-10), pp. 95–102 (2010)
3. Dal Pozzolo, A., Bontempi, G.: Adaptive machine learning for credit card fraud detection (2015)
4. Dal Pozzolo, A., Caelen, O., Waterschoot, S., Bontempi, G.: Racing for unbalanced methods selection. In: Yin, H., et al. (eds.) IDEAL 2013. LNCS, vol. 8206, pp. 24–31. Springer, Heidelberg (2013). https://doi.org/10.1007/978-3-642-41278-3_4
5. De Caigny, A., Coussement, K., De Bock, K.W.: A new hybrid classification algorithm for customer churn prediction based on logistic regression and decision trees. Eur. J. Oper. Res. **269**(2), 760–772 (2018). https://doi.org/10.1016/j.ejor.2018.02.009
6. Elazmeh, W., Japkowicz, N., Matwin, S.: Evaluating misclassifications in imbalanced data. In: Fürnkranz, J., Scheffer, T., Spiliopoulou, M. (eds.) ECML 2006. LNCS (LNAI), vol. 4212, pp. 126–137. Springer, Heidelberg (2006). https://doi.org/10.1007/11871842_16
7. Fisher, R.Å.: The Design of Experiments. Oliver and Boyd, Edinburgh, London (1937)
8. Good, P.: Permutation Tests: A Practical Guide to Resampling Methods for Testing Hypotheses. Springer, New York (2013). https://doi.org/10.1007/978-1-4757-2346-5
9. Gu, Q., Zhu, L., Cai, Z.: Evaluation measures of the classification performance of imbalanced data sets. In: Cai, Z., Li, Z., Kang, Z., Liu, Y. (eds.) ISICA 2009. CCIS, vol. 51, pp. 461–471. Springer, Heidelberg (2009). https://doi.org/10.1007/978-3-642-04962-0_53
10. Gutierrez, P., Gérardy, J.Y.: Causal inference and uplift modelling: a review of the literature. In: International Conference on Predictive Applications and APIs, pp. 1–13 (2017)
11. Hadden, J., Tiwari, A., Roy, R., Ruta, D.: Computer assisted customer churn management: state-of-the-art and future trends. Comput. Oper. Res. **34**(10), 2902–2917 (2007)
12. Idris, A., Khan, A.: Ensemble based efficient churn prediction model for telecom. In: 2014 12th International Conference on Frontiers of Information Technology (FIT), pp. 238–244 (2014). https://doi.org/10.1109/fit.2014.52
13. ITU: ITU releases 2018 global and regional ICT estimates (2018). https://www.itu.int/en/ITU-D/Statistics/Pages/stat/
14. Krieger, N., Davey Smith, G.: The tale wagged by the dag: broadening the scope of causal inference and explanation for epidemiology. Int. J. Epidemiol. **45**(6), 1787–1808 (2016)
15. Lemeire, J., Meganck, S., Cartella, F., Liu, T.: Conservative independence-based causal structure learning in absence of adjacency faithfulness. Int. J. Approx. Reason. **53**(9), 1305–1325 (2012)
16. Liu, X.Y., Wu, J., Zhou, Z.H.: Exploratory undersampling for class-imbalance learning. IEEE Trans. Syst. Man Cybern. Part B Cybern. **39**(2), 539–550 (2009). https://doi.org/10.1109/tsmcb.2008.2007853

17. Margaritis, D., Thrun, S.: Bayesian network induction via local neighborhoods. In: Advances in Neural Information Processing Systems, pp. 505–511 (2000)
18. Mitrović, S., Baesens, B., Lemahieu, W., De Weerdt, J.: On the operational efficiency of different feature types for telco Churn prediction. Eur. J. Oper. Res. **267**(3), 1141–1155 (2018). https://doi.org/10.1016/j.ejor.2017.12.015
19. Olsen, C., Meyer, P.E., Bontempi, G.: On the impact of entropy estimation on transcriptional regulatory network inference based on mutual information. EURASIP J. Bioinform. Syst. Biol. **2009**(1), 308959 (2008)
20. Pearl, J.: Causality: models, reasoning, and inference. IIE Trans. **34**(6), 583–589 (2002)
21. Petersen, M.L., Sinisi, S.E., van der Laan, M.J.: Estimation of direct causal effects. In: Epidemiology, pp. 276–284 (2006)
22. Raeder, T., Forman, G., Chawla, N.V.: Learning from imbalanced data: evaluation matters. In: Holmes, D.E., Jain, L.C. (eds.) Data Mining: Foundations and Intelligent Paradigms, pp. 315–331. Springer, Heidelberg (2012). https://doi.org/10.1007/978-3-642-23166-7_12
23. Scutari, M.: Learning Bayesian networks with the bnlearn R package. arXiv preprint arXiv:0908.3817 (2009)
24. Spirtes, P., Glymour, C.: An algorithm for fast recovery of sparse causal graphs. Soc. Sci. Comput. Rev. **9**(1), 62–72 (1991)
25. Spirtes, P., Glymour, C., Scheines, R.: Causation, Prediction, and Search, vol. 81. Springer, New York (1993). https://doi.org/10.1007/978-1-4612-2748-9
26. Tsamardinos, I., Aliferis, C.F., Statnikov, A.R., Statnikov, E.: Algorithms for large scale markov blanket discovery. In: FLAIRS Conference, vol. 2, pp. 376–380 (2003)
27. Verbeke, W., Dejaeger, K., Martens, D., Hur, J., Baesens, B.: New insights into churn prediction in the telecommunication sector: a profit driven data mining approach. Eur. J. Oper. Res. **218**(1), 211–229 (2012)
28. Verbeke, W., Martens, D., Baesens, B.: Social network analysis for customer churn prediction. Appl. Soft Comput. **14**, 431–446 (2014). https://doi.org/10.1016/j.asoc.2013.09.017
29. Zhu, B., Baesens, B., vanden Broucke, S.K., : An empirical comparison of techniques for the class imbalance problem in churn prediction. Inf. Sci. **408**, 84–99 (2017). https://doi.org/10.1016/j.ins.2017.04.015
30. Óskarsdóttir, M., Bravo, C., Verbeke, W., Sarraute, C., Baesens, B., Vanthienen, J.: Social network analytics for churn prediction in telco: model building, evaluation and network architecture. Expert Syst. Appl. **85**, 204–220 (2017). https://doi.org/10.1016/j.eswa.2017.05.028
31. Óskarsdóttir, M., Van Calster, T., Baesens, B., Lemahieu, W., Vanthienen, J.: Time series for early churn detection: Using similarity based classification for dynamic networks. Expert Syst. Appl. **106**, 55–65 (2018). https://doi.org/10.1016/j.eswa.2018.04.003

Author Index

Abecasis, Ana B.   34

Birattari, Mauro   3, 18
Bontempi, Gianluca   182

Caelen, Olivier   182

De Baets, Bernard   126
De Pauw, Joey   165
de Vries, Wietse   51
Dewitte, Jean-Christophe   182

Fanuel, M.   137

Garzón Ramos, David   18
Gharahighehi, Alireza   152
Goethals, Bart   165
Gomes, Perpetua   34
Goossens, Bart   85

Kegeleirs, Miquel   18
Kuckling, Jonas   3

Lebichot, Bertrand   182
Lenaerts, Tom   34
Libin, Pieter   34

Moens, Sandy   165

Nowé, Ann   34

Peck, Jonathan   85
Pérez-Fernández, Raúl   126

Saeys, Yvan   85
Schreurs, J.   137
Schreurs, Joachim   70
Spaey, Gaëtan   18
Suykens, J. A. K.   137
Suykens, Johan A. K.   70

Tang, Mengzi   126

Ubeda Arriaza, Keneth   3

Vens, Celine   152
Verhelst, Théo   182
Versbraegen, Nassim   34

Winant, David   70

Printed in the United States
By Bookmasters